BETT[illegible] SHORTWAVE RECEPTION

A Handbook for the Radio Amateur and Shortwave Listener

By William I. Orr, W6SAI

Radio Publications, Inc.
Wilton, Conn.

BETTER SHORTWAVE RECEPTION

Radio Publications, Inc.

Wilton, Conn.

Library of Congress Catalog No. 57-14916

First Edition

BETTER SHORTWAVE RECEPTION

TABLE OF CONTENTS

FOREWORD

A new world of thrills and excitement greets the adventurer who first sets foot into the realm of shortwave radio. Gone are the stereotyped programs and "short, transcribed announcements" that clog the dial of the broadcast receiver. In their place, a new world of intercontinental radio reception awaist the listener.

The hobby of shortwave reception is rapidly becoming more popular in this advancing world of science and invention. A decade ago, only a handful of shortwave enthusiasts followed a hobby that seemed far too complicated and expensive to Mr. Everyman. Today this misconception has been swept away, and shortwave radio takes its rightful place alongside other famous scientific hobbies such as "hot rods," "hi-fi," and photography.

As with these other hobbies, "know-how" is an invaluable tool to further the advance of the hobbyist. This book is presented in the hope that it will provide an introduction to shortwave radio for the beginner, and will present some of the practical aspects and helpful advice for the more experienced listener. Much of the information presented for the first time is the result of practical experience, and is given in the hope that it will save the reader the time-consuming process of learning it the "hard way."

Following the popular practice introduced in the "Novice and Technician Handbook," this Handbook offers easy-to-read, step-by-step assembly instructions for interesting, inexpensive receiving equipment, sure to be of interest to all shortwave listeners and amateurs.

The author wishes to thank the following individuals whose assistance and suggestions for the preparation of this Handbook were of great help:

George Jacobs, W3ASK
Bandel Linn, W4HXL
Wm. E. Mauzey, W6WWQ
B. A. Ontiveros, W6FFF
Wm. Vandermay, W7DET
C. F. Hadlock, W1CTW (National Co.)
Frank Lester, W2AMJ
(Hammarlund Mfg. Co.)
E. R. Mace, W8MWZ (Heath Co.)
E. Mullings, W8VPN (Heath Co.)
E. G. Schmeichel, W9YFV
(Chicago-Standard Transformer Corp.)
Kurt Wydler, HB9DS
(Swiss Broadcasting Service)
Floyd Backus, Newark News Radio Club
Oliver Ferrell
("Popular Electronics" Magazine)
J. Hummes (Miller Coil Co.)
J. M. Laporte
(Canadian Broadcasting Corp.)
A. H. Morgan
(National Bureau of Standards)
The Hallicrafters Co.
U.S. Information Agency

CHAPTER 1

Listen To The World

Come with me, away from the cares and worries of everyday living, on a trip of high adventure. Follow me down the electronic pathway of shortwave radio to far-off lands, to ships at sea, to explorers in the freezing outposts of civilization, and to radio amateurs conversing with the far corners of the earth. Hear the measured, stately tones of Big Ben in London marking the hours, or listen to the Kookaburra bird of Australia. Hear the satellites of Communism nervously defending themselves against the Voice of America. Tune across the airline channels, and the chattering radio teletypes pouring out the news of the day. Tarry awhile with a missionary in the Andes, speaking to another radio ham in Durban, South Africa, in cosmopolitan San Francisco, or in Dakar. You will find the world at your fingertips with a shortwave radio receiver!

The glamor and excitement of world-wide radio reception appeals to young and old alike. A satisfying and unusual hobby, it is the first step towards the radio amateur license. All that is required is a good shortwave receiver, a simple antenna, and basic knowledge of what you are looking for and where and when to look for it. Patience and practice will produce signals from all parts of the globe to delight the ear and excite the mind.

The Shortwave Spectrum

The shortwave region starts at the "bottom" of the standard broadcast band and extends to 30 megacycles or so. It has the unique property lacking in the staid and regulated broadcast band of propagating radio signals over the face of the planet for thousands of miles. Adventure and thrills are jam-packed in these international kilocycles. The purpose of this book is to enhance your enjoyment of this interesting and rewarding hobby of shortwave reception. Here's hoping the following chapters will give you better shortwave reception!

VOICE OF AMERICA speaks to the world in 41 languages over a network of 78 transmitters located in the United States and nine foreign countries. Three outlets (Munch, Germany; Luzon, Philippines; and Okinawa) have million-watt output power and are the world's most powerful broadcasting facilities. The network also includes a floating broadcasting station, the U.S. Coast Guard Cutter "Courier," now relaying broadcasts daily to the Communist world and free nations in the Near East from its anchorage off the Island of Rhodes in the Mediterranean Sea.

The Voice's headquarters are located in Washington, D. C., the point of origin for most of the daily programs. Shown above is the master control desk—the technical heart and brain of VOA. Selection of programs from 100 different sources for 26 separate transmissions may be made at one time. Direct lines to the White House, the United Nations, radio centers in Munich and Paris, and commercial networks permit up-to-the-minute selection of important news events and happenings in all parts of the free world.

The high frequency world is a score of times larger than the well-known broadcast band. Crammed into this vast range is a huge variety of odd and unusual forms of radio signals. A quick, preliminary spin across the dial of a shortwave receiver will provide a fascinating insight into the many uses of the high frequency spectrum.

Tuning the Bands

Bidding goodbye to the last broadcast station at 1600 kilocycles we tune into a new world, whose first inhabitants are the strong, steady signals of the police, marine, and navigation services. Cryptic messages as "*Car 102 go to the corner of Sixth and Pine. Code 16*" leave the casual listener with a vision of crime and intrigue barely duplicated by Dick Tracy. Sandwiched among the many municipal and state police radio stations is the high speed buzzing of the coastal LORAN *(Long Range Navigation)* stations, blanketing the oceans with position locating pulses for aircraft and ship navigation. Here, too, we find an occasional "160 meter" amateur, almost buried beneath the weight of Loran signals.

Slightly higher in frequency are the Marine Telephone operators, shepherding their loudly vocal fleet of fishing and pleasure craft. A sure sign of this service is the "buzz-buzz-buzz" of the telephone "busy signal" heard during listening periods. A few dial divisions away are spotted the boats, setting up a howl and din unbelievable to hear. Unlike the broadcast band, many shortwave stations operate on the same frequency channel, with only the strongest surviving! Listen a while and you'll soon know about the running of the Albacore, or when the *Sea Queen II* will dock at the pier.

As your dial slowly glides higher in frequency the pistol-shot signals of the radio-teletype stations appear, weaving a tapestry of world news at one hundred and fifty words per minute. Beside them is a ship-to-shore station calling the *Queen Mary*, or perhaps speaking to a disabled freighter, wallowing between black, tumbling waves. The nervous chatter of high speed

RADIO ANDORRA, broadcasting from tiny country high in the Pyrenees Mountains is rare DX catch for shortwave listener. One of the hundreds of stations received in the U.S., "Roc des Anelletas" transmits sponsored programs, contrasted to great majority of broadcasters that are government controlled.

commercial traffic blend with the musical, fading tone of a ship transmitter. What a world of riches is hidden in the mysteries of the International Code! Passing on, you suddenly hear *Concord Tower* at some unknown airbase giving landing instructions to a Navy flier approaching the runway with a smoking motor. A little higher in frequency *Flight 7173* is holding a long, involved conversation with a distant traffic controller regarding change of flight plan. The roar of the motors of *Flight 7173* can be heard behind the calm voice of the pilot.

At once you come upon a discordant squawking, and disconnected bits of code as you pass across the Novice portion of the "80 meter" ham band. A twist of the beat oscillator of the receiver turns the jumble into individual signals, painfully spelling out their story to other eager operators impatiently awaiting their turn to tell their fragmentary tale. Close to the frequencies of the beginners are the "phone" stations, busy with traffic nets and across-the-state-gossip. Searching the "phone band" will turn up mobile stations, the operators enjoying their hobby as they drive to the business of the day.

GIANT DIRECTIONAL ANTENNAS are used to transmit shortwave broadcasts to all points of the world. "Beamed" signal follows great circle route from transmitter to your receiver. The ionized layer of air above the earth reflects radio signals back to the earth, much like a mirror, permitting shortwave broadcasts to cover areas of many thousands of miles. Shown above is one of the many Voice of America's stations in the United States. The site of this "curtain array" is Delano, Calif., beaming programs to Asia and the Far East.

THE STRANGEST SHIP IN THE WORLD, U.S. Coast Guard Cutter "Courier" is anchored at the Island of Rhodes in the Mediterranean Sea. The vessel is a completely equipped radio station, including antennas, transmitters, generators, and other devices required for broadcasting purposes. A relay base for the VOA, the "Courier" beams programs to the Soviet Union and the Balkan-Near East area. The 150,000 watt medium-wave transmitter operates on 1259 kc, and the 35,000 watt shortwave transmitters operate on all shortwave broadcast bands.

In and about the jumble of voices and coded messages in the lower reaches of this spectrum are the first group of shortwave broadcast stations. The broadcast band of Latin-America cuts across this region, filling the air with eager voices of liquid Spanish. Castanets click and marimbas throb out the story of love unrequited. Some of the weaker stations are but a blur of sound, while others stand up and demand attention.

Another turn of the dial brings in a loud, musical tone, accompanied by a ticking clock, charting the passing seconds. Soon the tone pauses, as if for breath, and the station notifies you that you are attune to "*WWV, the National Bureau of Standards Station, Washington, D. C. When the tone returns, the time will be exactly . . . nine twenty-five p.m., eastern standard time.*" A pause, then the musical clock again chants its song, as the lighthouse of the air continues its never-ending watch of the hours. As you gain experience, you will soon find the Canadian Dominion Observatory Station, *CHU* at Ottawa, and the experienced listener can set his watch from the Tokyo time station, *JJY*.

Closely shouldering *WWV* are a jumbled group of aviation stations, point-

ONE HUNDRED AND SIXTY THOUSAND HAMS in over 260 countries of the world use the shortwave bands as their common meeting ground. Amateurs also operate "mobile" stations aboard ships and in aircraft. Many U.S. hams have complete stations in their automobiles. Above is the station of amateur Bob Merry.

to-point radios, and military stations, each intent upon its own affairs. Tune your dial at random in this region. You will hear a tow-barge conversing with the Navy base. Here is a central American station calling Miami, Florida. Intimately beside the Spanish accent is some unknown tongue, exhorting unseen listeners at an outlandish pace. The voice slowly fades away and a radioteletype signal flirts heavily with Brazzaville then departs. Look here! The Naval Radio Station is calling all ships of the "Western Sea Frontier" with the latest coded dispatch of the Admiral. Close by the Navy is some unintelligible monster, plucking tones unknown to earthly musicians from an ethereal instrument. What can it be? We pass on in wonder.

Shortwave Broadcasting

Once again you run across a burst of music. A pause reveals the *Voice of America,* with a message of freedom and hope. In the background is the heated snarl and buzz of an "Iron Curtain" jamming station, doing its best to obliterate the "Voice." As the message continues, the jammers cluster about as moths are attracted to a lamp on a summer night.

The next hundreds of kilocycles are a rich miscellany of world-wide broadcasting stations, occupying a frequency region as large as the standard broadcast band. The *British Broadcasting Corporation's North American Service* brings you the best in London and the Empire. The *Happy Station* of the Netherlands features the voice of Eddy Startz, known over the world for twenty years of broadcasting. The *Voice of the Andes* in Quito, Ecuador sends you religious programs and music. *The Call of the Orient* over *Radio Manila* features exotic music of the Far East. *Radio Ankara,* the *Voice of Germany,* the *Lighthouse of the Carribbean, Radio Brazzaville,* the *Voice of Zion;* one after another they cross the dial with music and news of the world we live in. The heavy-handed propaganda programs from Communist China, Bulgaria, Romania, and the Soviet Union are mixed with music and "news" of the latest twist of the party line. Days may be spent listening to the shortwave broadcast bands as new stations appear to take the place of some of the old. If you are lucky you might log a tantalizing bit of the clandestine "freedom stations" operating in parts of Soviet controlled territory.

The Amateur Bands

The hundreds of shortwave stations pass in a never-ending stream. At intervals across the dial the amateur bands appear. The local and domestic amateurs are loudly heard on the "80 meter" and "40 meter" bands, mixed with foreign amateurs on the "20," "15," and "10" meter bands. Patient combing of these spots will turn up almost all the countries of the world if

STANDARD FREQUENCY STATION WWV transmits on 2.5, 5.0, 10.0, 15.0, 20.0, and 25.0 mc. The accuracy of each frequency is better than one part in one hundred million. Signals are modulated with audio tones of 440 cycles and 600 cycles. One second time ticks are also given. Shown above is the vertical dipole antenna system of WWV located at Beltsville, Maryland, USA.

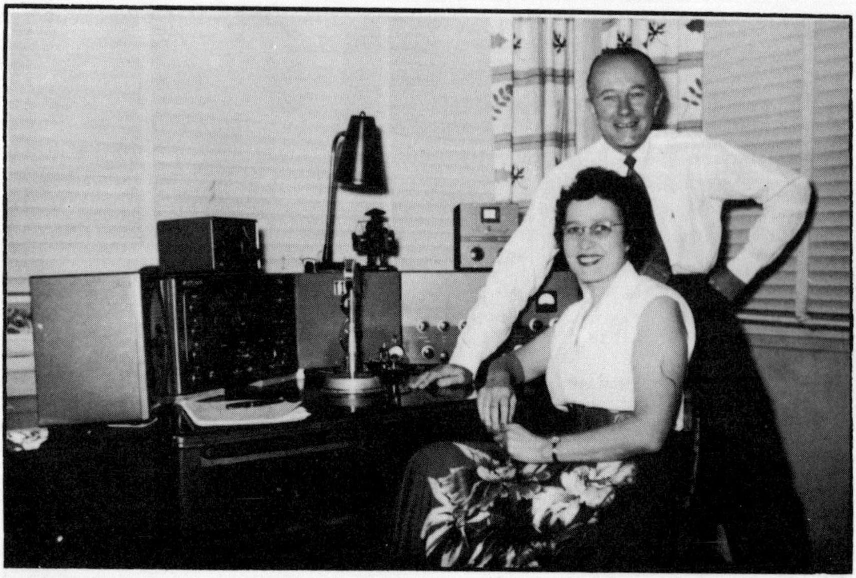

HUSBAND AND WIFE TEAM of Virginia and Ted Westlake operate amateur station TI2BX in San Jose, Costa Rica. Using a 120 watt transmitter, these Central American hams talk to their many friends in the U.S.A. and over the world. Many foreign countries permit operation of amateur stations by U.S. Armed Forces personnel. Greenland, Iceland, Germany, France, Morocco, Libya, Chile, Paraguay, Greece and Japan are represented by G.I. hams. Amateur radio is extremely popular in Japan which boasts of over 2900 native hams. Shown below is station of Kazatada Oohira, JA1CV, who operates on 14 mc and 21 mc phone.

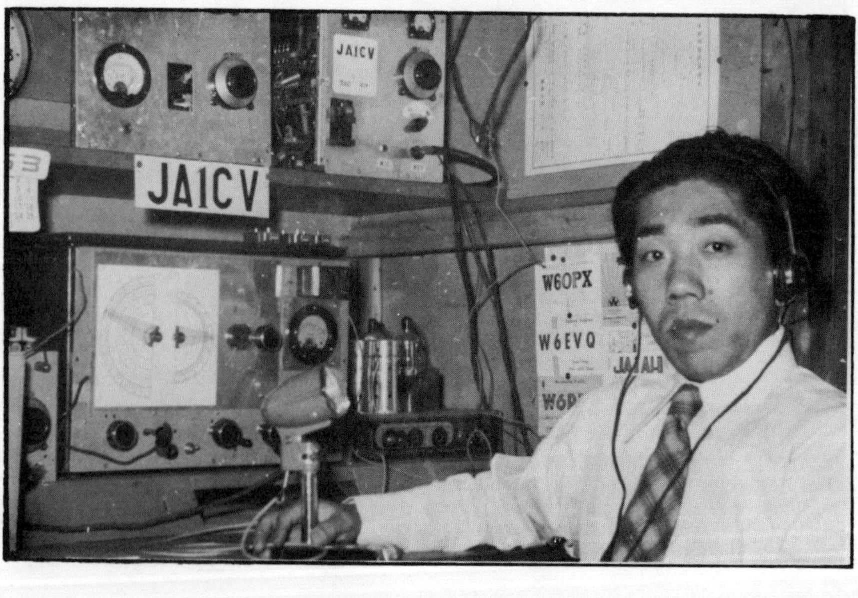

OUTSTANDING LISTENING POST of Floyd Backus, Richmond, Virginia, Floyd has logged over 200 countries on the shortwave bands. Three receivers are used: National NC-183, National NC-188. and RME-45. Antenna is 110 feet long, and 60 feet high, running east and west. Floyd is Vice President of the Newark News Radio Club, one of the oldest and most famous of listeners organizations.

you have a sensitive receiver and a good antenna. Let us listen to the "20 meter" band for a moment. Here is the powerful station of Swedish amateur SM5KP (operator "Vic") in Stockholm talking to the fading signal of a French expedition on Amsterdam Island. *(FB8ZZ)*. The location of the island? Better get out your World Atlas! Hear the clipped, Oxford tones of a British amateur talking to Upper Sandusky, Ohio, while an amateur in Florida copies messages he receives from the Byrd Station in Antarctica. *"It's*

GERMAN AMATEUR DJ2DW talks to American hams on the 10 meter band. Bert uses 35 watt Italian transmitter and beam antenna. His station is in Regensburg, Bavaria.

DELUXE HAM STATION of W6YMD, Pacific Palisades, Calif. is active on all high-frequency amateur bands. Specializing in DX contests, owner "Bill" Guimont has multi-operator installation known over the world. Antennas are at top of one hundred foot telephone pole placed at back of lot. W6YMD has six separate one-kilowatt transmitters and can switch from band to band in a few seconds.

only forty degrees below zero," says the operator at the South Pole. *"The cook is just starting to prepare a penguin egg omelette for our breakfast."*

At the opposite end of the earth amateurs in the wastes of the northern Arctic are heard speaking to the Thule air base and to Montreal. Just above them on the dial is an amateur in Rio de Janeiro. During the early morning hours the Australian and Far East stations appear on the dial. The powerful signal of *VS2DW* of Malaya can be heard beside the slightly weaker signal of *HS1A* in Bangkok, Siam. Here! Listen to this signal fading in and out! It is *XZ2AD* in Rangoon, Burma. He is talking to *4S7YL*, a YL operator of Ceylon. Try as we might we cannot receive the Ceylon station. Better luck tomorrow! In the American portion of the phone band a group of stations are excitedly calling *UA0KAD* in Siberia. Can we hear the Russian ham? Careful tuning brings in a fluttering signal with a deep voice, slowly speaking in English. A rare catch, indeed.

And so on through the day. Over 250 countries and islands have appeared on the amateur phone bands, plus another twenty or so on code *(c-w)*. Lucky indeed is the careful listener who has heard 'em all!

Above the "20 meter" amateur band lie more broadcasting stations, more

CANADIAN BROADCASTING CORPORATION, with studios in Montreal has three 50 kw shortwave transmitters at Sackville, N. B. One of the first s-w stations you will hear, "Radio Canada" broadcasts in sixteen foreign languages to Europe, South America, Australasia, and other areas. Above is Sackville station.

telegraphy and point-to-point services in an unending stream until the top of the tuning spectrum of the receiver is reached. Thrills? Yes, and excitement a-plenty, ranging from the calm, feminine voice directing the city prowl car on its journey of justice to the fading voice of an amateur station on the cold, windy plateau of central Asia . . . the voices of the great democracies with their programs of news and information for the world . . . the puppet-like signals of the Soviet-controlled Iron Curtain stations, extolling the delights of the police-state . . . the exotic voices from the odd corners of the world . . . *Radio International* of North Africa with native music strangely reminiscent of cats fighting on a fiddle . . . *Radio Andorra* playing jazz records with a Catalan accent . . . *All-India Radio* with the latest news of the democracies of the East . . . amateurs . . . *LA5YE*, Norway . . . *F8KW*, Paris . . . *SV1AD*, Athens . . . *ZS6ANE*, Johannesburg . . . *VU2ET*, India . . . *ZD8SC*, Ascension Island . . . *VP6FR*, Barbados . . . *KG6IG*, Iwo Jima . . . *CR9AH*, Macao . . . the voices of the world at your fingertips. These are the thrills and experiences awaiting the shortwave listeners! You can hear the world with a shortwave set!

CHAPTER II

Shortwave Reception—The Easy Way

To obtain the best results from your shortwave receiver it is a good idea to have some working knowledge of what radio is, and how long distance signals are propagated. This chapter explains the phenomena of long distance signal reception and gives you an insight into the operation of your receiver. The fine skill of tuning and logging shortwave stations is discussed, together with hints that will help your reception.

What is radio? This invisible, intangible servant of man is the greatest wonder of the world. Scientists can scrawl lengthy formulae on paper to explain the theory of the airplane, the secrets of the A-bomb, and the mathematical concept of hyperbolic, warped space. But of the ultimate nature of the radio wave, they know nothing. A simple definition of a radio wave is *electrical energy that has escaped into space. Radio,* in the overall sense, is not a *thing,* such as a chair or a house, but is a way in which things behave. Its effects can be measured and predicted, and it can be generated and put to use by man. Beyond that, radio falls into the twilight world of electromagnetic radiation which man is trying desperately to penetrate and understand.

Electromagnetic Radiation

Every electrical circuit that carries alternating current radiates a certain amount of this energy into space in the form of *electromagnetic waves.* The amount of radiated energy is small when the waves are large, as in the case of the ordinary 60-cycle electric light wires in your house. As the frequency of alternation is raised and the wavelength is lowered a region is reached where the invisible field becomes of some use for long distance communication. This region is termed the *communications range of the electromagnetic spectrum* ("radio waves"). Increasing the frequency of alternation above the range of radio waves leads first into the *infra-red region,* then into *light waves, ultra-violet waves, X-rays,* and finally *cosmic*

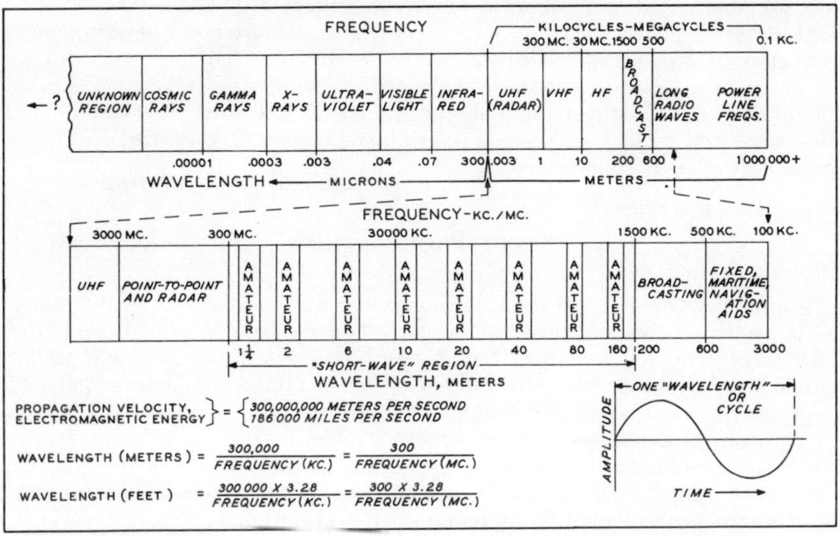

Fig. 1 Radio frequencies are but a small portion of the electromagnetic spectrum. As wavelength shortens, radio waves merge with infra-red waves. Waves may be described in terms of wavelength and frequency. The radio wave travels through space at a speed of 186,000 miles per second, or 300 million meters per second. The frequency, wavelength, and the speed of the radio wave are all related to one another as shown in the above equations. Use these equations to make conversions between frequency and meters!

rays. The whole gamut of radiation of all types falls neatly into place in the electromagnetic spectrum, as shown in Figure 1. Of great interest to the shortwave enthusiast is that portion of the spectrum falling between the higher limit of the broadcast band (1600 kc) and approximately 30,000 kc, for in this region extremely long distance radio communication takes place.

Before we examine this region more closely, let's pause a moment and examine some rather common terms that define the particular spot in the spectrum at which we may choose to tune our receiver. These terms are: *kilocycle, megacycle* and *meter*. Accompanying them are the terms *wavelength* and *frequency*. What do they mean?

Wavelength and Frequency

A confusing situation exists in that any given radio signal may be described in terms of *frequency* (rate of change of alternation of the wave) or in terms of *wavelength* (physical length of one wave cycle). The radio wave (like all electromagnetic energy) travels through space at the speed of light which is about 186,000 miles per second, or 300 million meters per second. The meter is a unit of measure of the physical sciences and is approximately 3.28 feet in length. Because the radio wave occupies both time and space the wavelength and the frequency and the speed of light may all be packaged into a neat little formula that will translate wavelength into frequency and vice-versa, as shown in Figure 1.

As a quick mental exercise, let's assume a transmitter is operating on a frequency of 25,000 kilocycles (25 mc). By dividing the speed of light by the frequency, we find that the wavelength of the station is 12 meters. You will note that the conversion data of Figure 1 will work for meters and kilocycles, and meters and megacycles. Just be sure to keep track of your zeros, and all will come out well!

Radio Propagation

Although we cannot see the electromagnetic radiations leaving the transmitting antenna, instruments can measure and describe it, and theory and experiments can tell us what it does after it has left the antenna. Radio signals leaving the antenna are influenced by a number of factors, some of which may be controlled, and others which are beyond our power to control. We can control the angle and the direction taken by the signal as it escapes into space by making changes in the electrical characteristics of the antenna and by changing the height of the antenna above the surface of the ground, but after the wave has left the antenna system we are powerless to control it.

A portion of the radiated energy hugs the surface of the earth and is termed the *ground wave*. Commercial broadcasting stations make use of the ground wave, since it has proven to be a very reliable means of communicating over medium distances. Ground wave coverage in the broadcast band may range up to 500 miles, the coverage increasing as the frequency diminishes. Large portions of the earth may be covered by the ground wave at frequencies in the 100 to 200 kilocycle region. At the higher frequencies

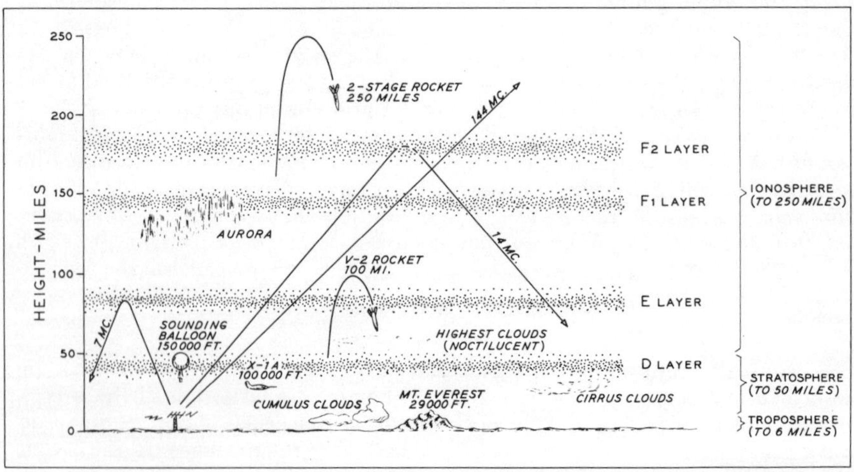

Fig. 2 The earth's atmosphere is composed of three layers, the lowest of which is the troposphere, or weather layer. The stratosphere (constant temperature zone) is next, extending to a height of about 40 miles. Above this is the ionosphere (ion layer) which is the region of reflection of radio signals.

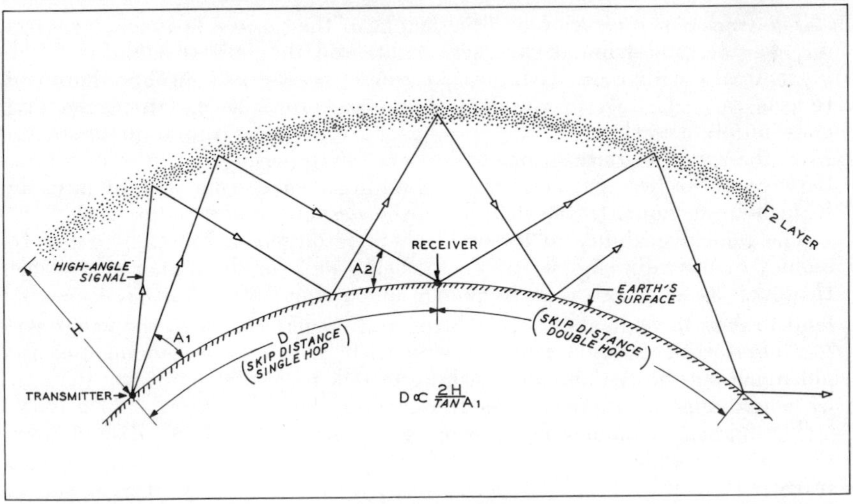

Fig. 3 Single hop transmission takes place up to distances of about 2500 miles. Beyond this the angle of radiation becomes too low to clear the earth, and radio signals must reach more distant areas by means of multiple-hop transmission. Silent zones (skip distance areas) between the hop points may be noted. Occasionally, the signal may reach a distant point by travelling the "long path," opposite in bearing to the great circle path by 180 degrees. Signals arriving over both long and short paths have bad echo and flutter.

(shorter wavelengths) the ground wave diminishes, dropping to a dozen miles or so in the vicinity of 10 mc. The ground wave does not again become useful until the VHF (Very High Frequency) region above 50 mc is reached. Ground wave propagation is used at VHF for television, FM, and other forms of propagation covering limited areas of high population density.

A second portion of the radiated energy is termed the *sky wave* which increases in effectivity as the frequency of operation is increased. First noted in the early "twenties," the sky wave permits communication at great distances that are not normally reached by the high frequency ground wave. An additional characteristic of the sky wave is that it skips over the earth, jumping large areas and then showing up once more beyond the area of *skip distance*.

After a good deal of head scratching, the savants decided that the sky wave type of propagation was brought about by an area of the upper atmosphere known as the *ionosphere* (Figure 2). This interesting region hovers between 60 and 150 miles above the surface of the earth and "bounces" high frequency radio signals back to earth that would otherwise be lost in outer space. In this region of the upper atmosphere, energy from the sun strikes particles of the air and ionizes them, releasing free electrons from the particles. The blanket of electrons formed by this action has the unique ability to reflect radio signals of certain frequencies back to the surface of our planet. Radio signals aimed at this invisible mirror are reflected back to earth, many thousands of miles away from the transmitter.

The degree of ionization and the height of the ionized layer changes from day to day, and from darkness to light, and the reflective ability of the layer varies with these changes. At times the layer is in fine fettle and reflects radio signals in robust fashion, and all is well. Often, the layer goes mildly beserk from the lashings of ultraviolet radiation emanating from the sun, and radio transmission via ionospheric skip "goes to pieces." Between these two extremes radio conditions vary from day to day, and from hour to hour, to the delight or the despair of the radio ham.

The reflective ability of the ionosphere is largely a function of the frequency of the radio signal, in general being best for the higher frequencies. However. as we progress in frequency much above 30 mc, the radio signals tend to pass through the ionized layer and do not return to the earth. This is a comforting thought to those who wish to talk to the moon, but it is not much help to the amateur wishing to talk to distant points on the earth by employing the higher radio frequencies.

The highest frequency reflected back to earth between two distant points may be termed the *maximum useable frequency* (MUF) for that particular transmission path. The MUF seems to be a function of the 11 year sunspot cycle, the highest values of MUF being recorded at peaks of the cycle.

Ionospheric Variations

Nothing is still in nature. The earth, moon, and sun all change their positions with respect to one another and the electromagnetic radiations from the sun wax and wane. Short wave transmission conditions fluctuate

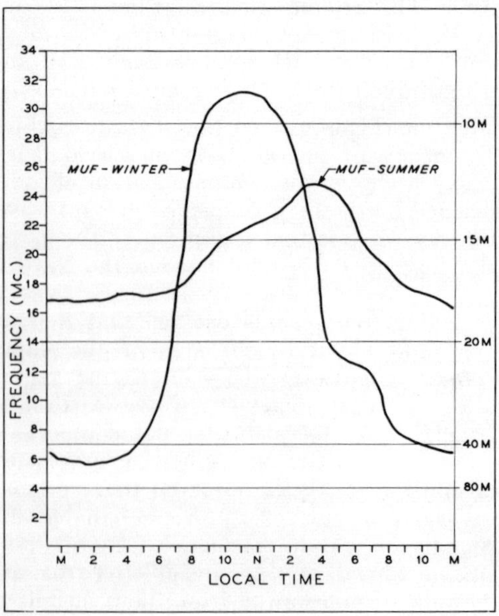

Fig. 4 The MUF for a typical east-west path during the low period of the sunspot cycle rarely peaks over 17 mc. At the top of the cycle the MUF may rise to 50 mc or more.

TREMENDOUS ANTENNAS and high power transmitters are used by major shortwave broadcasting stations to overcome vagaries of ionospheric transmission. Shown at right is technician making adjustments on transmitting tower of VOA relay base at Munich, Germany.

from day to day, and from hour to hour. Over the years, a study of propagation conditions has clarified the situation, permitting forecasts to be made in advance of coming radio conditions. Like the weather, however, the radio conditions often do not follow the "advice of the weatherman," but long term variations in propagation can be computed, and the picture of the never-ending fluctuations becomes clearer.

Strongly ionized layers of the atmosphere reflect radio waves of high frequency that would easily pass through weakly ionized layers. Thus propagation conditions are dependent upon the degree of ionization. Ionization decreases when the earth swings past the orbital range farthest from the sun, and radio conditions may be expected to deteriorate during these summer months. Because the axis of the earth is tilted from the perpendicular to its orbit the winter nights are long compared to the twilight hours of summertime. Both these properties have a profound effect upon shortwave radio conditions. The extended winter periods of darkness allow greater time for the various layers of the ionosphere to de-ionize and to pass larger chunks of the radio spectrum into space.

The effect of these fluctuations in the ionosphere set a maximum frequency limit of reception for a given path (MUF) which varies from season to season as shown in Figure 4. The MUF for an east-west path across the United States is shown for summer and winter conditions. During the long summer days when the earth is at the outermost reaches of its orbit, ionization of the upper air is weak, but relatively constant over a 24 hour period, and the MUF varies over a restricted range of frequencies. Communication is impossible on paths whose propagation figure lies above the maximum useable frequency, and the path gradually decreases in effectiveness as the operating frequency is decreased below the MUF.

The winter days bring the earth closer to the sun, increasing the level of ultra-violet radiation falling on the ionosphere. The long winter nights allow a longer period of de-ionization. As a consequence the MUF reacts in a more violent manner, dropping to low levels at night, and then increas-

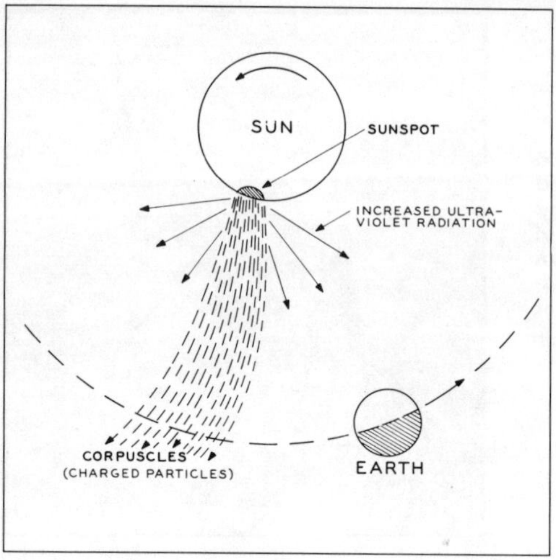

Fig. 5 Sunspot storm bombards earth with charged corpuscles creating ionospheric chaos. The MUF "radio dropout" is sometimes avoided by the earth passing ahead of corpuscular cloud.

ing to a maximum figure during the hours of the day. Thus, during a winter evening, a noticeable thinning out of stations will be apparent as the higher frequencies pass across the dial. Signals from these stations are passing through the ionosphere, escaping reflection to ears of listeners on earth.

At intermediate frequencies, communication may be supported between distant points almost 24 hours a day. This comes about when the frequency is close to the MUF. As shown in Figure 4, a transmitter operating on 16 mc could maintain communication over this particular path with the exception of a few questionable hours around 4 p.m., local time, when the MUF would be too high for good, reliable communication. Highly directional antennas and large power levels help the major shortwave broadcasting stations make the best of the situation when the qualities of the ionosphere are marginal. For the reader who is interested in pursuing this topic further, additional reference material is given at the end of the book. As a result of the undulations and changes in the radio mirror above the earth, shortwave reception does not occur on the steady, reliable basis we are accustomed to expect from broadcast band stations. The shortwave signal that breaks your eardrums may be a whisper in a few hours, and completely gone the next day. The station that can barely be heard at noon may be of overpowering strength at five p.m. This only adds to "the sport of the chase" and makes life more interesting to the avid shortwave listener!

During certain years the sun has large prominences on the surface that may be seen by looking at it through a piece of dark glass. These *sunspots* are of unknown origin, proof of violent agitation on and within the sun. Occasional huge solar flare-ups or *sunspot storms* are apparently connected with sharp increases in ultra-violet radiation and result in the emission of electrified particles or corpuscles which bombard the atmosphere of the earth. The increase in radiation from the sun is responsible for *sudden ionospheric disturbances* (SID's) while the corpuscular particles following at a

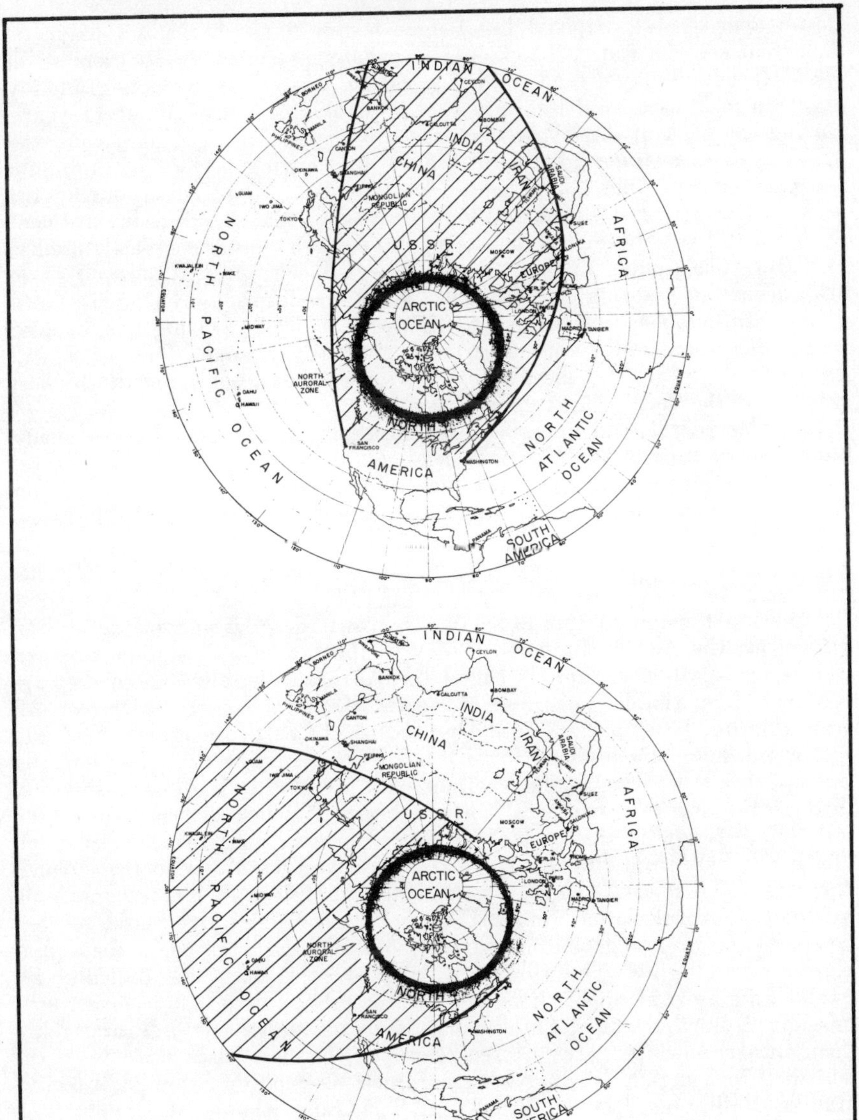

Fig. 6 An arctic projection showing the approximate location of the northern auroral zone. Cross-hatched area shows part of world that cannot be effectively reached by shortwave from U.S. due to shielding effect of the zone. Note that circuits from New York to Tangier and from San Francisco to Manila are not shielded by this zone. Below is shown the auroral shielding of transmissions from Tangier. Note that areas shielded from New York can be covered by relay station in Tangier. Same effect is true of relay base in Manila.

slower speed create delayed but longer lasting ionospheric and magnetic disturbances. The radio "black-out" is usually preceded by an increase in the MUF which may extend as high as 50 mc or so. The sunspots gradually build up in number and intensity over a regular period of about 11 years. During the peak of this cycle when the solar spots are most numerous, the MUF tends to average quite a bit higher than during periods of minimum sunspot activity. Thus radio conditions tend to become better during the peak of the 11 year sunspot cycle, even though the erratic bombardment of the earth by corpuscular emissions from the sun creates a greater number of radio "black-outs" during the same period. The present sunspot cycle peak occurred in spring of 1957, and the cycle will gradually decrease until a new minimum is reached some time about 1963. During the coming years, therefore, radio conditions will gradually change. The MUF will drop year by year, and the exodus of shortwave stations from the highest reaches of the radio spectrum will begin.

In order to take advantage of the best propagation conditions, many international broadcasting stations transmit the same program on several different frequencies, taking advantage of the various different MUF's on difficult propagation paths, such as those near the North and South Poles.

The Auroral Zone

Covering the top of the globe like a crown lies the *auroral zone,* best known for the Aurora Borealis display in the heavens. A similar ring encircles the south pole, and is called the Aurora Australis. These displays are caused by visible ionization of clouds of charged particles thrown off from the sun. These particles are captured by the earth's magnetic field and follow the magnetic lines of force toward the geomagnetic poles, ionizing the upper atmosphere as they collide with its atoms. Radio transmission through the auroral zone is difficult since the efficiency of reflection of the ionosphere is low in this area. The ionized crown, therefore, blocks much of the Far East from effective radio reception and transmission from North America (Figure 6). Signals passing through this area are recognizable by their "auroral flutter" created by uneven reflection from the ionized layers of the area. The flutter may be so rapid that it renders the signal almost unintelligible. The *Voice of America* overcomes this difficulty by establishing auroral zone by-pass stations, or relay points that receive transmissions from the United States, boost them in signal strength, and relay them into the desired area of reception. *VOA* relay bases have been established in Tangier, North Africa; Munich, Germany; Salonika, Greece; Manila, Philippine Islands; Honolulu, T.H.; the islands of Rhodes and Okinawa; Colombo, Ceylon; and England. Most all of these relay stations place strong signals into the United States and they may be heard with ease.

The Great Circle Route

Radio transmission between two points may be considered to follow the shortest path consistent with its reflections back and forth between the earth and the ionosphere. This path is known as the *great circle route.* This route is the shortest distance between two points on the surface of a

sphere, and usually appears as a curve on the regular maps. Great circle maps can be made for a specific locality, and several are shown at the rear of this Handbook for various areas of the United States. The great circle map is unusual in that a point of origin is chosen, and is placed at the center of the map. All directions radiate from this point, and a ruler laid from the center point to any other point on the map automatically traces a great circle route. Such a map is indespensible if it is desired to "aim" a directional antenna at some point thousands of miles away.

How To Tune A Shortwave Receiver

Oh, yes! There's an art to tuning a shortwave receiver! Even though the shortwave spectrum is broken into three or four bands on most sets, each band covers a tremendous amount of territory. Roaming at random across the dial may be easy, but it will not produce satisfactory results if you wish to hear distant stations! The first step in the right direction is to learn the purpose and function of all the knobs on the receiver. There is an ironic joke among receiver engineers that the instruction manual of the receiver is something to be thrown away with the packing box! They reveal that most of the letters written by receiver owners to the manufacturers ask questions that are fully covered in the receiver instruction manual. Read it! Practice tuning the receiver and use all the controls until you understand the many functions of the set.

When you have the "feel" of the receiver, try your first serious listening on the lowest frequency shortwave band. This covers the smallest segment of the spectrum. The signals are also stronger on this band and exhibit less fading than those higher in frequency. Tune the main frequency dial *slowly* and note how the stations come into audibility and fade out again. Note also that the stations are spaced closely together at some parts of the dial, and are farther apart at other places. Observe that each station occupies a much smaller fraction of the dial scale than does a broadcast station. Practice tuning in areas of interference and try separating one station from the other. A few minutes of practice in carefully tuning signals will pay big dividends.

Practically all shortwave receivers incorporate *automatic volume control* to reduce the effects of fading and auroral flutter. Always make sure that you use the a-v-c system when you listen to a telephone signal. Sometimes interference can be reduced by manipulation of the audio tone control. A high pitched hetrodyne between two adjacent stations can be reduced by turning the control towards bass reception. Under most circumstances the tone control should be advanced towards the treble end to allow faithful reproduction of speech, and to prevent the signal from sounding "boomy." Experienced shortwave listeners, incidentally, do most of their DX-chasing with earphones, rather than with the loudspeaker. Reception and comprehension are much better when external room noises are barred from the ear.

Next, turn the bandswitch to a high frequency band and practice tuning in a foreign broadcast station. Notice that it is harder to tune in a weak, fluttering signal than it is a strong, local one. Then too, it takes plenty of listening practice to enable you to understand words transmitted via a fluttery signal. Plenty of listening practice will boost your comprehension of weak and fluttery signals. It has been said that some amateurs interested in DX can place a pillow over their heads and hear two wires scratched together in Outer Mongolia! With plenty of practice, you can be that good, too!

C-W Reception

The next phase of listening is to master code (c-w) reception. C-w may be sent with a hand operated key, a semi-automatic key, or by tele-typewriter. The simplest form of code transmission is for the action of the key to turn the radio transmitter on and off. This does not make much of a difference in the receiver unless the *beat frequency oscillator* of the receiver is turned on. The b-f-o creates a beat note, or audible hetrodyne with the c-w signal, resulting in musical code signals that may be heard with ease. For best code reception the a-v-c system of the receiver should be turned off, and the r-f gain control is used to adjust the intensity of the signal. The audio gain control is turned *full on*, and the r-f gain control is adjusted for optimum signal strength. The beat oscillator control is adjusted to provide a pleasing tone or pitch to the c-w signal. Experiment with different settings of the b-f-o pitch control to find the one best suited to your taste. Always operate the receiver with the audio control full on and the r-f gain control retarded.

Single Sideband Reception

Many amateurs and commercial stations make use of single sideband

(SSB) transmission, wherein the radio carrier and one sideband are suppressed and only one sideband is transmitted. To the uninitiated, these signals bear a great resemblance to the squawking of a duck! It is necessary to restore the carrier to the signal before it becomes intelligible. The simplest method of carrier restoration is to use the b-f-o of the receiver as a local carrier. The receiver is set up as for c-w reception with the audio gain advanced and the r-f gain retarded. The b-f-o is *off*. The single sideband station is tuned in for maximum signal strength. Leaving the tuning dial alone, the b-f-o is turned on, and the pitch control of the b-f-o is *slowly* tuned back and forth. The squawking of the duck will gradually change pitch and start to resemble human speech. After the b-f-o control is adjusted for best intelligibility, the main tuning dial of the receiver may be retuned *very slightly* for best audio quality. Admittedly, this is a "cut-and-try" system of tuning a SSB signal, but unless your receiver has a calibrated b-f-o and you know which sideband the station is transmitting, this tuning sequence is as good as any other. Plenty of practice will do wonders to enable you to tune in SSB stations quickly and easily.

CHAPTER 3

How To Buy A Shortwave Receiver

If you are a Texas Oil Millionaire the best way to buy a good shortwave communications receiver is to shine the glare from your diamond stick-pin in the eye of the salesman in the radio store, gently tap the ashes from your five dollar seegar into his shirt pocket, and say, "Son, wrap up a dozen of your most expensive receivers. I'll take 'em all. One of 'em is bound to be good!"

Since there may be one or two readers who are not Texas Millionaires, the following points are offered in the hope that they may be of assistance to the prospective purchaser of a new receiver.

A receiver is pretty much like an automobile. The basic motor remains almost the same from year to year as the body style changes. Some developments are announced with a fanfare and then fade into oblivion—like "free wheeling." Like the automobile, receivers come in all sizes and prices, and some of them are good and some of them are not so good. And, like the automobile, the receiver grows old, starts to "burn oil" (blow out components), and is finally traded in on a new model. To carry the comparison to the final conclusion, there exists good buys in second hand receivers, just as can be found on the used car lot. This idea will be pursued later in this chapter. Right now, we'll lift the lid and peek into one of the brand new receivers and try and determine its true value.

The Skeleton

The stripped bones of a communication receiver are illustrated in Figure 1A. A converter tube, one i-f stage, a combination detector and audio stage, an output stage, and a power supply total five tubes. The receiver usually covers 550 kc to 30 mc in four tuning ranges. In most instances it is an ac-dc set, employing no power transformer and having the filaments of the tubes in series across the power line. It probably has a built-in speaker. The cost of such a set is fifty to sixty dollars. Despite the few tubes and the lack of an r-f stage this "stripped down" receiver will give

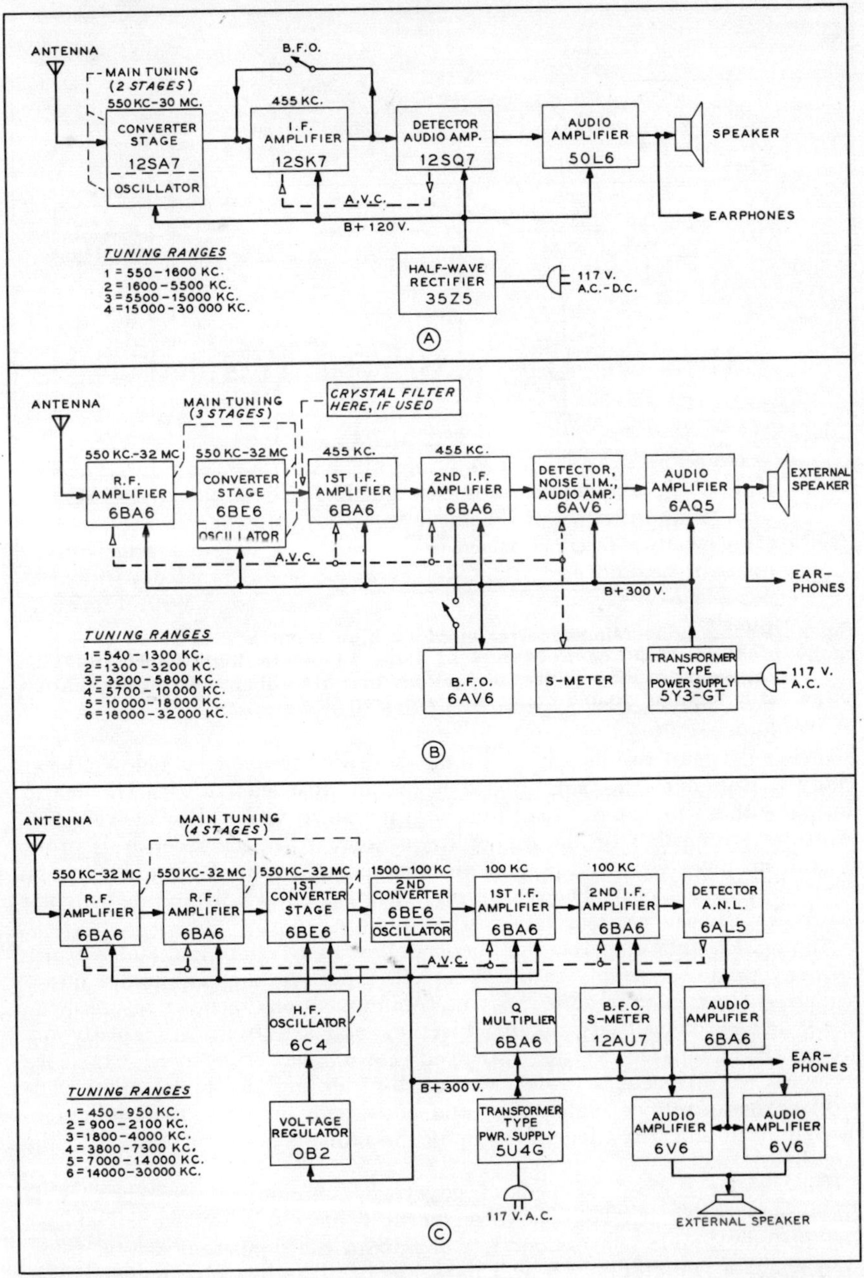

Fig. 1 "Basic" shortwave receiver is shown in (A). Addition of r-f stage, beat oscillator, and crystal filter (B) produces utility communication receiver. Deluxe receiver having two r-f stages is shown in (C). Note use of 100 kc i-f stages.

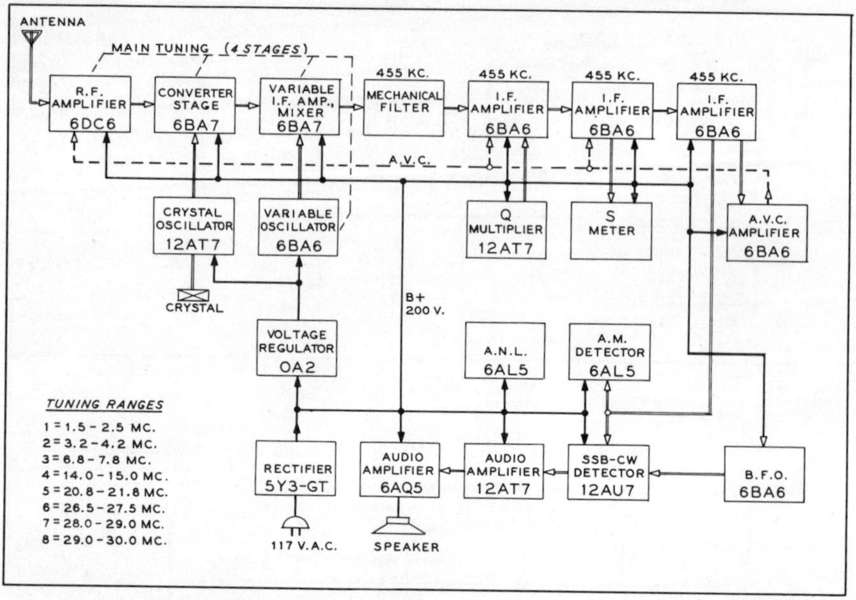

Fig. 2 Double conversion receiver employs high frequency i-f stage to reduce image problem, and low frequency i-f stage to obtain high selectivity. First conversion oscillator may be crystal controlled for highest stability. Collins 75A-4 (page 33) is popular double conversion receiver.

surprisingly good results, especially on the lower frequencies below 10 mc. Many a beginner has "cut his eye teeth" on just such a set. The major defects with a simple design are that both the sensitivity and image rejection of the receiver suffer on the higher frequencies. Also, the receiver is prone to pick up power line noises that are coupled into the r-f circuits through the ac-dc power supply. The set, however is a good buy for the money, since many ac-dc broadcast receivers cost as much.

The next higher cost receiver incorporates an r-f amplifier and a transformer-type power supply in the design. In addition, a noise limiter and a second i-f stage may be added. A true communications receiver results from these changes. Sensitivity, image rejection, and selectivity are greatly improved as compared to the "stripped down" set. The more expensive receivers in this category also add an S-meter and a crystal filter. The receiver has seven or eight tubes and costs from one hundred dollars to two hundred and fifty, depending upon the refinements incorporated in the circuitry.

The "Rolls Royce" of general coverage communications receivers is diagramed in Figure 1C. The receiver may have one or two r-f stages, and include double conversion. It will have a multi-position crystal filter, or perhaps a Q-multiplier. It will have plenty of audio, a regulated power supply, and an extremely rugged chassis. Critical circuits will be air-tuned and temperature compensated. A receiver of this class will cost from three hundred dollars to over six hundred dollars. It is strictly for the "pro-

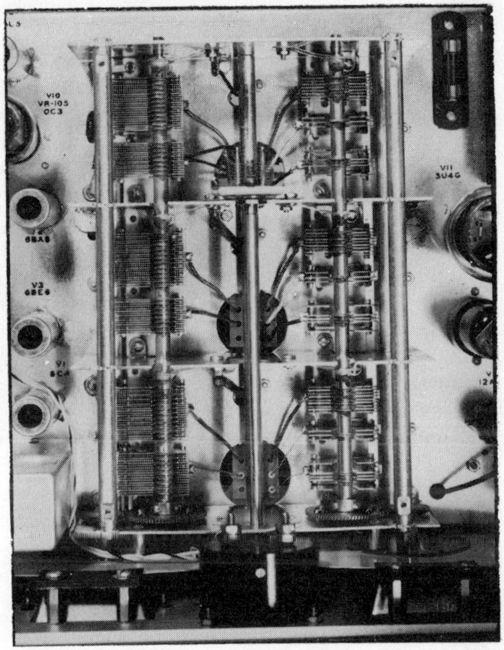

TUNING ASSEMBLY of six band receiver uses multi-section variable capacitors to obtain proper tuning ratio. Main tuning gang is at left, and bandspread sections are at right. Coil catacomb is placed directly below the chassis. (Hammarlund HQ-140XA is shown in photograph.)

fessional," since it requires a practiced hand to obtain all the benefits from its complicated circuitry. It may have as many as eighteen tubes (and almost as many knobs!).

A species to itself is the strictly "amateur communications receiver." A receiver of this type covers only the narrow segments of the spectrum occupied by the amateur bands, to the exclusion of the greater portion of the short wave region. It is designed to do a tough, specific job, and it does it well. It will not serve as a general receiver for casual listening, as it is designed and built for a particular type of service. For the serious amateur desirous of the best this type of receiver represents the ultimate in perfection. The block diagram of such a receiver is shown in Figure 2.

What to Look For When You Buy a New Receiver

The first thing to decide is "what brand of car" you wish to buy. In most cases, this decision is automatically made by the pocketbook. The real choice must be made between various receivers falling within the same price class. Excluding the basic receiver of Figure 1A, the following discussion applies to all other grades of receivers.

Stability—Sensitivity—Selectivity

These three terms can be used to sum up the performance of any receiver. In order that you may evaluate the performance of your receiver, or one that you intend to purchase, an explanation of these terms is in order.

Stability

The electrical stability of the receiver is a function of the *thermal* stability and the *mechanical* stability. The thermal stability is a measure of the response of the receiver to changes in temperature. As the receiver warms up, the heat of the tubes and the power transformer gradually warm other components in the receiver. The tuned circuits are particularly sensitive to temperature changes. The thermal stability of the receiver is the ability of the receiver to stay tuned to a station after a short warm-up period. As the heat increases in the receiver cabinet, especially in the vicinity of the conversion oscillator, component values will change, causing the dial setting to slowly vary. This effect is known as *drift,* and can be very annoying if it is necessary to keep retuning a station for several minutes or even an hour after the receiver is first turned on. All receivers, except the most expensive have varying amounts of thermal drift, and it is well to check this drift before a new receiver is purchased.

Mechanical stability is a function of receiver design. The converter stage is especially sensitive to vibration in most receivers. Instability may be caused by loose components, poor tubes, or wires that move slightly when the receiver is vibrated. Many receivers having the speaker mounted in the case are prone to mechanical instability when the set is playing loudly. The vibrations of the speaker can lead to acoustic feedback, taking the form of a loud howl heard on strong signals. A well designed high frequency receiver will have a heavy case, with all parts firmly mounted to the chassis. Heavy leads will be employed in the r-f section of the receiver to reduce effects of mechanical vibration. Rubber shock mounts are employed in some cases to reduce vibration of the sensitive tuned circuits.

Sensitivity

Sensitivity is the ability of the receiver to receive weak stations over the inherent noise of the set. Of the three requirements for a good receiver, this one is the easiest for the manufacturer to meet. All but the most inexpensive receivers (having no r-f stage) have sufficient sensitivity and amplification to "pull in the weak ones." Generally speaking, the more tubes a receiver has, the greater gain it has. However, the amount of gain often does not really indicate just how sensitive the receiver really is. Some poor receivers having excessive gain merely amplify their own noise!

Selectivity

Selectivity is the ability of a receiver to reject unwanted stations. The true criteria of a good receiver is not what its receives, but what it rejects! This is the most critical requirement of the three, and the one where most receivers fall flat on their fuses. Because it is impossible to design a tuned circuit that will pass only one frequency, stations on either side of the wanted one will usually be heard, and will cause interference. Generally speaking again, the lower the intermediate frequency, and the more i-f stages, the better will be the selectivity of the receiver.

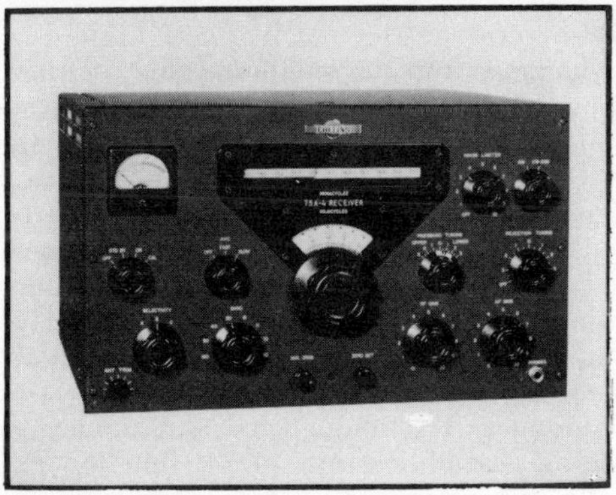

COLLINS 75A-4 is 8-band ham double conversion receiver. Covering 1-megacycle slices of the spectrum, this receiver offers maximum in way of image rejection and selectivity. Mechanical filters are used in the i-f amplifier.

"Four Band" and "Six Band" Receivers

After hanging around the local radio store for a while, you will noice that some receivers cover the 550 kc to 30 mc spectrum in four ranges, while other receivers cover the same chunk of spectrum in six ranges. You can obtain this information merely by looking at the bandswitch of the receiver and noting the number of positions on it. Further cogitation upon this subject will reveal that—in general—the inexpensive receivers have four bands, and the more costly ones have six bands.

An obvious question is: Why are two different bandswitch systems used on shortwave receivers? Why don't all receivers either have four bands or six bands? How-come??? The answer to this query touches upon the economy of building a shortwave receiver in production quantities, and upon the desired signal-to-noise ratio of the receiver. It is easy to see that the least costly receiver would be one that had only one tuning range. This would do away with the bandswitch and the extra coils. The tuning dial of this receiver would cover 550 kc to 30 mc in one enormous range. If such a receiver was to be made, great economies would be gained, since a whole collection of expensive parts could be eliminated. A receiver of this type, unhappily, would be impractical for several reasons. First of all, the dial would tune over 24,450 kilocycles—almost twenty times as great a range as the whole broadcast band! Each shortwave station would occupy only an infinitely small speck on the dial. In order to tune a particular station, an expensive gear reduction system would have to be attached to the tuning capacitor of this receiver. In addition, the bearings of the capacitor would probably have to be hand-made and of the finest quality. The money saved by eliminating the bandswitch and extra coils would therefore be lost in the expensive dial mechanism necessary to tune the receiver.

Even if it were possible to manufacture an inexpensive high precision tuning dial, a second problem effectively eliminates the "single band" concept. The "fly in the ointment" centers around the action of the tuned circuits of the receiver when an excessive frequency span is desired. The coupling efficiency and impedance of the tuned circuits are in proportion to the ratio of inductance and capacitance required to tune the receiver to a given frequency. Other factors being equal, the sensitivity and signal-to-noise ratio of the receiver are highest when the largest value of tuning inductance and the smallest value of tuning capacity are used. The useable sensitivity of most receivers gradually drops off as the capacity of the tuning capacitor is increased. You will notice, therefore, that most amateur bands are placed at the high frequency end of the tuning dial of your receiver. Sensitivity and receiver performance is best at this tuning point, since the tuning capacitor gang is almost fully open. This deterioration is caused not only by a drop of tuned circuit impedance, but also by the fact that it becomes harder to keep all circuits properly aligned as the tuning capacity is varied. The greater the variance in capacity, the more complicated the alignment problem. This difficulty *could* be overcome as it is largely a problem of economics. If enough time and money was spent upon making precision tuning assemblies they might function exceedingly well over an enormous frequency range.

This whole philosophy of design, however, is headed in the wrong direction. Receivers have been built in the past which attempted to tune great segments of the shortwave spectrum without resorting to switching mechanisms. They proved to be expensive and unsatisfactory. The builders regretfully junked the idea of a single band receiver and turned their thoughts to systems of bandswitching.

The Tuning Ratio

The key to the problem seemed to be in obtaining acceptable results with a minimum number of tuning ranges. Experiments proved that a *tuning ratio* of about 2½ to 1 is a good compromise between economy and proper receiver operation. The tuning ratio is defined by the ratio of the of the highest tunable frequency on a given band to the lowest frequency. Thus, if a receiver tunes from 6mc to 18 mc, the tuning ratio is 18/6, or 3/1. If the receiver tunes from 6 mc to 12 mc, the tuning ratio is 12/6, or 2/1. The tuning ratio of the broadcast band is about 2½/1, so a variable capacity designed to tune over the broadcast frequencies will also work well covering the same tuning ratio on the shortwave frequencies. The use of a common capacitor for both shortwave and broadcast tuning is an economy that helps to allow receiver manufacturers to produce excellent, low cost shortwave receivers. Early receivers having this tuning ratio worked well over the shortwave spectrum, and it shortly became standard practice to hold close to this tuning ratio in most receivers. Only four different sets of coils are required to cover frequencies up to 30 mc or so with a 2½/1 tuning ratio, and the "four band" shortwave receiver became a reality.

The only disadvantages to the four band receiver were that the over-all sensitivity of the receiver tended to drop off at the low frequency end of the tuning range, and that alignment of the r-f circuitry became a bit critical. These difficulties could be overcome by improved design and assembly techniques of the four band coil assembly, or by the use of six bands instead of four. Both techniques are in use in modern receivers. The six band receiver covers the 550 kc to 30 mc range, with tuning ratios of 1.8/1. The sensitivity of the receiver is more constant over the smaller tuning range, and the alignment problems seem less severe. This tuning ratio

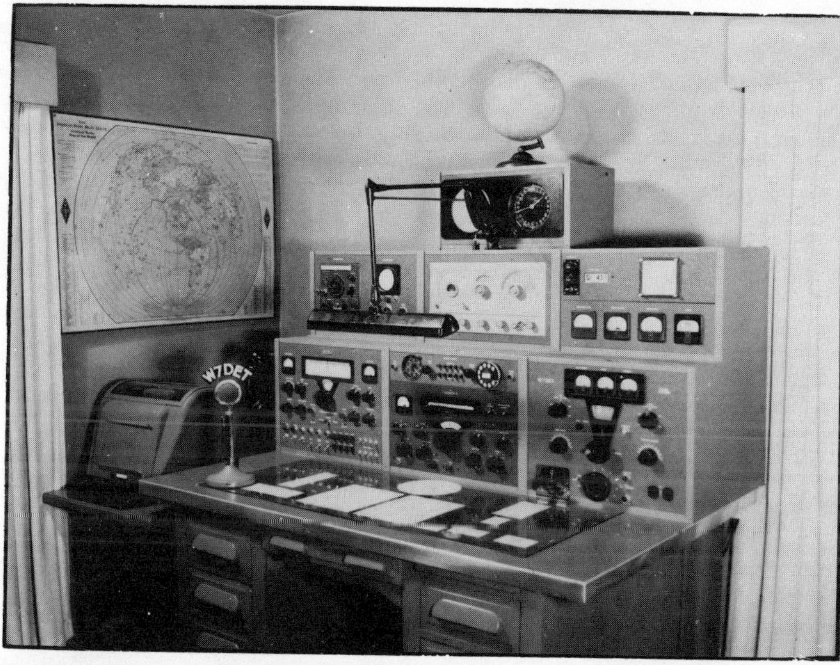

RADIO W7DET OF SEATTLE, WASHINGTON is outstanding example of home built equipment. Complete station fits atop a standard desk. Transmitter exciter is at left, with all-band variable frequency oscillator. Kilowatt all-band amplifier is at right. Control panel is at center, over Collins 75A-4 receiver. Across top are control equipment, auxiliary receiver, and clock.

is employed in the *Hammarlund HQ-140X*. A receiver having an 1.8/1 tuning ratio costs more than one with a 2½/1 tuning ratio, since extra coils and bandswitch segments are required. The receiver performance, in general, is improved over the simpler "four bander," but at an increase in complexity and cost of the r-f circuitry.

It should be emphasized that either type of design works well. A good four band receiver will "run rings around" a poor six band receiver. You pay your money and you take your choice! As you well know, more Volkswagen autos than Lincoln Continentals are sold in the world! This is because the Volkswagen is less expensive than the Continental, and more people wish to own Volkswagens than Continentals. Either car will take you from one place to another. The one you buy depends upon your taste in motor cars and your pocketbook! If you are the famous Texas Oil Millionaire, you should buy the flashiest and most expensive six band receiver you can find. The rest of us will have to use our best judgment and pick and choose. In some instances the smaller car is better suited to the particular job than is the more expensive one. So it is with receivers. You can own a "Volkswagen" receiver or a "Continental" receiver. The choice is up to you.

How to Buy a Second Hand Receiver

The watchword for buying a second hand receiver is *caveat emptor*—let the buyer beware! You can pick up a dandy bargain in a second hand receiver, or you can get "caught with a lemon." Unfortunately in some cases the "lemon" looks good, and the bargain receiver may have a kicked-around appearance. The problem, then, is to be able to judge whether the receiver is good or bad before you pay your money for it!

So let's assume that you have a few bucks burning a hole in your wallet, and you want to buy a good short wave receiver. What to do? Should you pick up a brand new model and pay the full price, or should you buy a second hand receiver that someone else has used (perhaps not wisely but too well). This choice must be left up to the individual. As in the case of a second hand car, some people like 'em, and some people don't. If you are not adverse to the idea of a used receiver, the following hints will help you to pick out a good one.

First of all, you must *see* and *hear* the receiver before you buy it. Buying a receiver through a classified ad, or from a person in a distant town is a risky business. His idea of a "good, clean receiver" may be a lot different than yours is! Also, it is not easy to get any kind of a guarantee as to receiver operation in a situation like this. The best bet is to purchase the receiver from a store or from an individual that will give you an opportunity to try the receiver out before you buy it.

Following this advice, let's walk to the nearby radio distributor who has a whole room full of second hand receivers. Maybe we can find a good buy! After explaining to the salesman that we're shopping for a second hand receiver, he waves us into a small room with eight or ten receivers sitting on a long bench. "Help yourself, boys," he states. "Try 'em all, and let me know which one you want."

Armed with this permission, we look around the room. Not bad! The ten receivers represent eight different types, ranging in price from thirty to four hundred and fifty dollars. The first problem is to decide which one we are interested in. Having set a maximum limit of one hundred bucks, let's see what we can buy for that amount of money.

The salesman poked his head in the door. "Why don't you try that Super-Blooper 100 in the corner. Cost $278 new. You can have it for $140." A little rich for the blood, but a good place to start. Let's go down a "check list" and see how the receiver measures up to it.

Check List for a Used Receiver

1—Turn on the receiver and adjust it to the broadcast band. Listen to several broadcast stations. Make sure the S-meter (if any) works. It should "kick up" in a lively fashion. If it does not, or if the voice and music sounds muffled it is a warning that perhaps the automatic volume control section of the receiver is inoperative. Paper capacitors are often used in the a-v-c circuit, especially in receivers that are over five years old. These capacitors deteriorate with age, and the a-v-c action of the receiver gradually "goes to pot." Check the receiver for a-c hum. If it is excessive the filter capacitors are probably bad.

2—If the receiver seems to function well on the broadcast band, switch to each of the short wave bands in turn. You should be able to receive signals on each band if a good antenna is attached to the set. Notice if the background noise level is the same at each end of the dial. If the background noise level is greatest at the high frequency end of a band, it means that the alignment of the r-f circuitry for that particular band is not in adjustment.

3—Check the dial calibration of each band. Look for Standard Frequency Station WWV, broadcasting a musical tone and time-ticks on 2.5, 5, 10, 15, 20, and 25 mc. Observe how much error in dial reading occurs when the receiver is tuned to this station. Also, tune in the amateur bands and see if they occupy the proper place on the dial. If WWV shows up at the wrong place on the dial, don't waste your money calling the National Bureau of Standards to get them back on frequency, since the receiver is probably out of calibration.

4—If the receiver passes these tests, it is time to find out some of the more subtle points that differentiate a "good buy" from a "lemon." Tune in a strong station on the highest frequency band. Tune it in carefully and let the receiver sit at that dial setting for ten minutes or so. If the station gradually drifts out of tune, the receiver is temperature sensitive. Unfortunately, most receivers are more or less susceptible to this form of thermal drift, so you should not judge the receiver too harshly in this respect. However, the tuning should "settle down" and remain relatively constant 30 minutes or so after the receiver has been turned on. If a continual drift over a period of hours is noted, you had better pick out a different receiver!

5—The heart of the receiver is the band change switch. A good band change switch will last for hundreds of thousands of operations. Once in a while, a switch will be found with dirty or defective contacts. If you buy a receiver with a bad band change switch, you have "had it," since it is an expensive, major operation to replace it. To test the bandswitch, tune in a loud station on the highest frequency band. Carefully flip the bandswitch to the next lower frequency band, and then return it to the high frequency band. Can you still hear the original signal? No? Then turn the tuning dial and hunt for it. If the receiver requires retuning to the original signal when the bandswitch has been turned and reset it is an indication of dirty or intermittent contacts in the bandswitch assembly.

This is a rather critical test, and many new, modern receivers right out of the packing case will not pass it. It is an expensive task to make a receiver assembly absolutely rigid, and the action of turning the band change switch can and does warp the chassis a small amount. This, in turn alters the configuration of the high frequency oscillator of the receiver, changing the dial setting a few kilocycles.

However, you can tell a lot about a used receiver from the bandswitch. If the tuning returns to within a kilocycle or two of where it was before you made the test, and no excessive noise develops when you rock the band change switch back and forth on the contact, the switch is probably in acceptable shape. If the receiver breaks into loud noises when the bandswitch is turned, or the original signal is found several hundred kilocycles down the dial from its original position, the operation of the band change mechanism is open to suspicion.

6—Tune the receiver to the high frequency end of the highest frequency band. Advance the audio volume control and listen to the background noise of the receiver. Remove the antenna and see if there is a distinct drop in background noise, indicating the internal noise level of the set is less than the external atmospheric noise. If the receiver you are interested in has no r-f stage (that is, it has a two gang rather than a three gang tuning capacitor) you can forget this test, as the receiver is not sensitive enough to differentiate between the two types of noise. A good receiver should generate little noise when the antenna is removed.

7—The receiver should be checked for images. An *image,* as explained later in this Handbook, is a "repeat" signal from a loud station. That is, the station is heard at a second point on the dial. All receivers have images to a greater or lesser degree. You should make sure that the images of your prospective purchase are not excessive. If the receiver in question has no r-f stage, you might as well forget this test, since the image rejection of this type of receiver is very poor. That is one price you must pay in an economy receiver.

Tune the receiver to the 10 meter or 15 meter band and hunt for a strong signal. Note the frequency of the signal, then look higher or lower in frequency for a "repeat" or image signal. If the receiver is a single-conversion job having an intermediate frequency of 455 kc (ask the salesman about the intermediate frequency), the image signal

QSL CARDS GALORE! Voice of America received over 5300 reception reports from 60 countries during the month of January, 1957 for the program "Panorama, USA." Mrs. Helen MacKenzie, clerk, of the English Language Branch of VOA adds more mail to the pile.

will be either 910 kilocycles above or below the main signal. In most instances it will be lower in frequency than the test signal. The amount of attenuation of the image signal will give you a good idea of the image rejection of the receiver. If the receiver is properly aligned, there should be a definite difference in signal strength between the desired signal and the image. The greater the difference in strength, the better is the image rejection capabilities of the receiver. If there is little or no difference between the two signals, the receiver is either incapable of good image rejection, or it is out of alignment. In either case, the situation should be explored further before you "lay down your long green" for the receiver.

If two receivers of the same make and model are available for test, try both of them on the same signal and see which one has a better image rejection. Also, try receivers of different makes on the same signal. If you can hear WWV on 20 mc, it will make a good, steady signal for test purposes. You will require a good S9-plus signal to make this test, otherwise the image signal might be too weak to readily find.

8—Next, test the auxiliary controls to see if they are in good shape. Make sure the volume control and the r-f gain control are not noisy in operation. If the set has a crystal filter, be sure that it works. Check the tone control for operation. See that the pilot lamps light.

9—At this stage of the game, see if the owner of the set or the store salesman will remove the bottom plate of the receiver so you can examine the components. Look for charred spots on resistors that would indicate a short circuit at some time in the past. Examine the bypass capacitors. If they are tubular units, observe if the wax has been forced out of the ends of the capacitor, a sure sign of a leaky unit. Above all, see if the previous owner has incorporated any changes in the circuitry. If additional wiring or components that do not belong in the set are found, you had better "tread lightly," as you have no idea what unusual modifications the previous owner might have made.

10—Finally, make sure that you obtain an instruction book with the receiver. Attempting to align a receiver, or to trouble-hunt without the assistance of an instruction manual is like washing your feet with your socks on. You can do it, but it isn't fun!

So here you are with your second hand receiver picked out! Chances are it will not measure up to 100% on the above checkout list. If you are a "whiz" on receivers you can take it home and probably correct all the little complaints. If you are not an expert, or do not have the suitable equipment to repair and align the receiver you might be able to have the store align it for a small charge. Do not assume that if the receiver is up for sale it has been completely aligned and checked out. Many radio dealers take in old receivers in trade on new ones, and they consider the old ones to be "pains in the neck," since the mark-up on used receivers is small. As a consequence not much time is spent getting the old receiver into tip-top shape.

Finally, remember that "you get what you pay for." The very inexpensive receivers—the new ones—are little more than toys to begin with. When you buy a second hand one you are usually buying a pile of junk, as the receiver has no margin of safety to carry it through years of service. The older, more expensive receivers can be bought for 30% to 60% of their new price. Therein lies the bargains, for a good, old receiver is many times better than a new "cheapie." The only deterrent is that *you* are the one who has to find out which old receiver is good, and which is unsuitable. Here's hoping that this information will be of help to you.

Build Your Own Receiver?

The manly art of building your own receiver is almost as extinct as the famous dodo bird. A decade or so ago the home-made receiver was the rule rather than the exception. The main obstacle to receiver construction is that the total cost of parts required to build a good receiver is almost as high as the cost of a completely manufactured unit. In addition the "trade-in" value of a home-made receiver is just about zero. Nobody wants some other man's construction project. On the other hand, you can always recoup from 50% to 75% of the original cost of the manufactured receiver when you get ready to dispose of it. As with automobiles, some receivers depreciate faster than others. One of the most famous and expensive receivers depreciates only about 10% a year.

Of great interest to the enthusiast with a thin pocketbook is the fact that a commercial receiver may be bought through many radio stores on a time payment plan. Although interest is usually charged in such a case, the receiver can be enjoyed and used while you are paying for it. To carry home and use a $250 dollar receiver for a down payment of fifty dollars or so is a luxury that was impossible not many years ago.

CHAPTER 4

Know Your Receiver!

Are you a real "DX-hound" or merely a "knob tweaker"? Do you know how your receiver really works, or do you merely treat it as a piece of fine furniture? Look inside your receiver, and also remove the bottom plate. Do you see all those trimming capacitors and variable adjustments? When they are properly aligned your receiver will have ears like an Iroquois hunting party. When these adjustments are not properly made, your receiver will be as dead as a Scotch hamlet on Tag-Day. You can make minor adjustments to your receiver and make sure it is in tip-top condition if you know the purpose of all these little controls. The purpose of this chapter is to discuss these adjustments and how they affect receiver operation.

Any receiver, no matter how expensive, usually requires a checkup after it has been used for a period of time. You don't have to "change the oil every 1000 miles" as you do with your automobile, but you should have a periodic checkup to insure that your receiver will always perform to its greatest capability. The very minimum you should do is to clean the dust and dirt from the interior of the receiver every six months or so, and take the tubes out and check them in a good tube tester. A small ten-cent store paint brush will do a fine job of removing dust from the corners of the receiver chassis.

As soon as you get the receiver you should hook it up to your antenna and jot down notes concerning the performance in a notebook. You should log the error in dial calibration at each end of each band, make a note of the normal background noise, and a notation of the setting of the audio volume control for normal listening volume. If the sensitivity of the receiver should gradually decrease over a long period of time you would notice it as it would be necessary to gradually advance the setting of the volume control with the passage of time.

Someday the time will arrive when you decide that it is time to realign the stages of the receiver for maximum sensitivity and proper calibration.

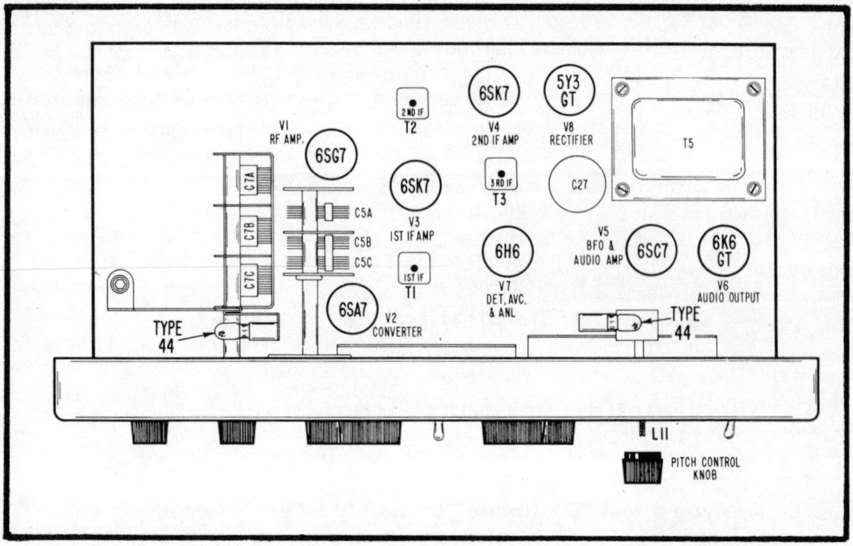

Fig. 1 Parts layout of typical shortwave receiver is logically arranged to permit shortest possible leads in radio frequency section. R-f amplifier and converter tube are placed adjacent to tuning capacitor, with bandswitch and coils directly below the chassis. The i-f amplifier is at the center.

You can do this the easy (and the expensive) way by taking the set back to the store from which it was purchased. You can save money and learn a lot, however, by aligning the receiver yourself.

Receiver Circuitry

Before you touch the "innards" of the receiver, it is a smart idea to familiarize yourself with the schematic of the receiver, and to correlate the part notations on the schematic with the actual components of the receiver. Remove the bottom dust cover of the receiver and examine the circuitry. Take a pencil and mark on the chassis near each tube socket the tube designation and its function—i.e.: 6SG7, r-f amplifier, etc. That will help you to keep things straight while you trace out the circuit.

If your receiver is a true communications set, it will have a three gang tuning capacitor attached to the main dial, and perhaps a second three gang capacitor attached to the bandspread dial. In general, the following observations will apply to most all receivers of this class. First, the capacitor section nearest the main dial is the oscillator tuning capacitor. This unit determines the spot in the radio spectrum to which the receiver is tuned. The middle capacitor section is the mixer tuning capacitor, which tunes the signal grid circuit of the mixer tube. The rear capacitor section is the r-f amplifier tuning capacitor. If the set employs metal tubes, the r-f amplifier tube is usually a remote cutoff pentode such as the 6SK7, or 6SG7. The mixer may be a 6SA7 or perhaps a 6SB7-Y. Some older receivers used a

6K8 mixer tube. In some instances the mixer tube also performs as the conversion oscillator, thus even though the tuning capacitor of the receiver has three gangs, only two tubes are required to do the work of three. If a separate oscillator tube is used, it is probably a 6J5 triode. The oscillator tube is usually closest to the front panel, the mixer tube immediately behind it, and the r-f amplifier tube is near the rear of the three-gang tuning capacitor. A general receiver layout is shown in Figure 1.

If miniature tubes are used, the choice of tubes is somewhat greater. A 6BA6, 6DC6, 6BZ6, or 6CB6 may be employed as the r-f amplifier. The mixer tube is usually a 6BE6, with a 6BA7 and a 6U8 or 6X8 running a slow second. Once in a while a 6AU6 or a 6AK5 is used as the mixer tube. In the less expensive receivers, the pentagrid mixer is also used as the conversion oscillator, while the more expensive receiver employs a separate oscillator tube, such as a 6C4. By noting the position of the tubes nearest the tuning capacitor it is easy to determine which ones comprise the r-f section of the receiver.

Most communication receivers employ one or two intermediate frequency stages, which usually operate at a frequency of about 455 kc. The 6BA6 is the most popular miniature tube for this service, while the 6SK7 is used in "metal tubed" receivers. Occasionally a 6BD6 miniature tube will be used in the i-f circuit for reasons best known to the receiver manufacturer.

The second detector and audio sections of the receiver vary greatly from one manufacturer to another, and from one model to another of the same

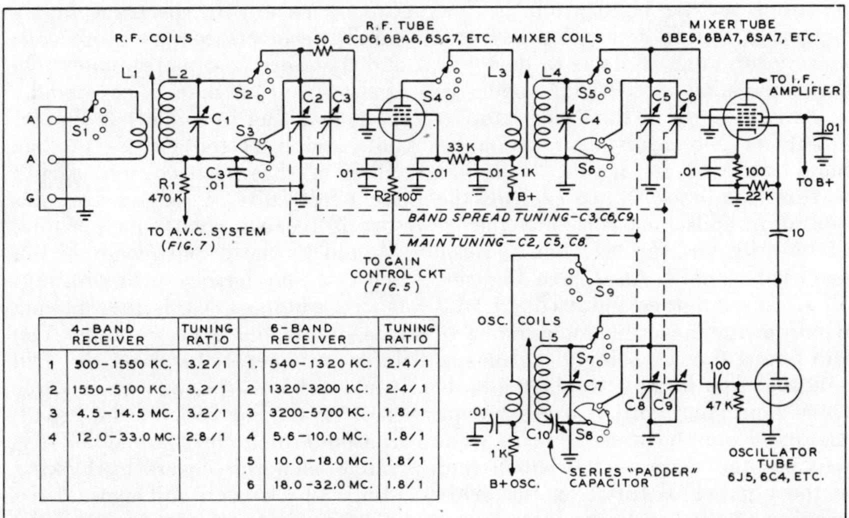

	4-BAND RECEIVER	TUNING RATIO		6-BAND RECEIVER	TUNING RATIO
1	500–1550 KC.	3.2/1	1.	540–1320 KC.	2.4/1
2	1550–5100 KC.	3.2/1	2	1320–3200 KC.	2.4/1
3	4.5–14.5 MC.	3.2/1	3	3200–5700 KC.	1.8/1
4	12.0–33.0 MC.	2.8/1	4	5.6–10.0 MC.	1.8/1
			5	10.0–18.0 MC.	1.8/1
			6	18.0–32.0 MC.	1.8/1

Fig. 2 Coil switching is employed in majority of shortwave receivers. Unused coils are shorted out by S3, S6, and S8. Oscillator and mixer functions may be combined in one dual purpose tube. Proper tracking of the oscillator and r-f circuits is accomplished by series "padder" capacitor C10 in oscillator. The series capacitor determines the rate of tuning of the oscillator which is slightly different than the tuning rate of the mixer and r-f circuits.

manufacturer. In general, a 6AL5 or a 6H6 (metal) may be used as a second detector, a-v-c, and noise limiter followed by one or two stages of audio amplification using 6AQ5 or 6V6 (metal) tubes.

The more expensive receivers may incorporate extra a-v-c tubes, S-meter tubes, and double conversion circuits to cloud the issue. Even so, armed with the schematic, and with the receiver upended before you, you should soon be able to pick out the major circuit components.

The Radio Frequency Section

You will note that the r-f coils mounted below deck are arranged around the bandswitch in such a manner that the interconnecting leads from the coils to the switch, and from the switch to the tube sockets are as short as possible.

Once you have determined which set of coils belongs to the oscillator, and which sets belong to the mixer and r-f amplifier respectively, you are ready to study the r-f circuitry more closely. The schematic of Figure 2 is a general circuit that is common to most receivers having an r-f stage. Compare it with the schematic of your receiver. Trace out the wiring between the switch assembly, the tuning capacitors, and the tubes on a separate piece of paper and compare it with Figure 2 and with the schematic in the instruction manual. Let's take each stage at a time and go over the circuit.

Most receivers cover the broadcast and shortwave range with four positions on the bandswitch. A few receivers cover the spectrum in six ranges. It can be deduced that the "six band" receivers require more coils and more switch contacts to do the job and therefore cost more money. On the other hand, each range covers less of the spectrum than is covered by any single range of the "four band" receiver, resulting in somewhat simpler alignment and adjustment problems. A discussion of the virtues of "six band" receivers as against "four band" receivers has been covered earlier in this Handbook. Since each of the three r-f circuits of the receiver has one set of coils for each band, the "four band" receiver should have a total of 12 coils, and the "six band" receiver should be the proud owner of not less than 18 coils. As seen in Figure 2, each coil may have several windings on it. The r-f stage input coil, L1-L2 has two windings. L1 is the antenna winding and has a small number of turns. L2 is the grid winding. You will notice that each input coil has a different number of turns on the grid winding. The highest frequency coil has only a few turns, and each lower frequency band coil has a greater number of turns. The coil for the broadcast band may have many turns wound in a *bank* or *honeycomb* style. It is possible, therefore, to tell which tuning range each coil covers by looking at the number of turns on the grid winding. Check each coil against the position of the band switch knob, and the position of the wiper arm of the switch.

It is common practice to short out the windings of the unused coils in each stage to prevent them from reacting with the active coils. The coil switch (S2-S3) will therefore have two decks. Deck S2 selects the proper coil, and deck S3 shorts out the other coils. For economy, these two decks may be mounted on opposite sides of one switch segment.

VOICE AND CODE automatic announcing equipment at Bureau of Standards station WWV. Accurate frequency transmission of this station is useful for the alignment and adjustment procedure required by all shortwave receivers.

The mixer coils (L3-L4) are very similar to the r-f input coils, as they cover the same frequency range. Upon close examination, it can be observed that the primary winding (L3) has more turns upon it than does the corresponding primary winding (L1) of the input coil. This is necessary to achieve good signal transfer from the r-f tube to the mixer tube. In all probability, grid winding L2 of the input coil will be exactly the same as grid winding L4 of the mixer coil. Three separate switches are employed to choose the proper mixer coil: Deck S4 selects the plate coil for the r-f tube, deck S5 selects the proper grid coil for the mixer tube, and deck S6 shorts out the unused grid coils. These three decks may be mounted upon two switch segments. Occasionally, you will find a receiver where (by a burst of ingenuity) all three switches are mounted upon one segment! A veritable rat's nest of wires will lead to this one deck!

Bandspread Tuning

All modern short wave receivers employ *bandspread tuning* whereby a small portion of the high frequency spectrum may be spread over a large segment of the dial for ease of tuning and logging accuracy. The bandspread may be a mechanical type, using a gearing system which permits a slow rate of tuning with an auxiliary tuning knob. This corresponds to "low gear" in an automobile, where the rear wheels turn slower than normal at any given motor speed. In the case of the receiver, the tuning capacitor turns slower than normal for any given speed of rotation of the bandspread tuning knob. One dial and two tuning knobs may be employed, or a separate bandspread dial and tuning knob may be added to the receiver. In either case, only one three section tuning capacitor is used. This type of bandspread is commonly used in inexpensive receivers having no r-f stage. It is not costly and it works well.

More expensive receivers employ electrical bandspread which requires a second three gang tuning capacitor. This capacitor may be readily recognized since it will have only two or three plates per section. In some receivers it is incorporated as part of the main tuning capacitor. A separate dial and tuning knob is used to tune the bandspread capacitor. In Figure 2, the main tuning capacitor gang is C2-C5-C8, and the bandspread capacitor gang is C3-C6-C9. The bandspread dial is usually calibrated in kilocycles for each amateur band. In addition it often has a 0-100 degree calibration for general use at other portions of the short wave spectrum.

When electrical bandspread is not desired the bandspread dial is set so that its capacitor gang is at minimum capacity (plates unmeshed). When it is desired to use the bandspread dial, the main tuning dial is set to the *high frequency* side of the portion of the spectrum over which bandspread action is desired. The bandspread capacitor is then slowly meshed, adding circuit capacity and tuning the receiver lower in frequency. Thus all bandspread action normally starts from the high frequency end of the bandspread dial and progresses lower in frequency. Let's assume that you want to bandspread the 14 mc amateur band. The main tuning dial is set slightly above the high frequency end of the band, say to 14.4 mc with the bandspread dial set to its maximum indicated frequency (capacitor plates unmeshed). Tuning the bandspread dial will now cover the 14 mc band, starting at the high frequency end and advancing toward the low frequency (14.0 mc) end.

Padding Capacitors and Slug Tuned Coils

If a short wave receiver was precision-made by an electronic engineer who measured each component to insure its absolute accuracy, and who aligned the receiver with the most expensive and precise instruments, it would be a fine set, indeed. It would also cost about $10,000. It is a tribute to the receiver manufacturer that he can approach this stage of excellence on a production basis and yet sell the receiver for a modest sum. In order to do this, he has to have some means of compensating for the normal, slight variations inherent in all radio tubes and components. To insure that the tuned circuits of the receiver cover the range they are intended to, padding capacitors and slug tuned coils are employed, permitting close adjustment of the circuit parameters.

To begin with, the minimum amount of circuit capacity must be determined, for even with the main tuning capacitor and the bandspread capacitor set at zero (plates unmeshed) the wiring, components and tubes contribute a measureable amount of residual capacity that must be taken into account. This "bottom" capacity may wander over a wide range, and varies with make of tubes used, placement of parts, lengths of interconnecting leads, and slight tolerance variations in manufacture of the individual components. To establish a standard value of "bottom" or minimum capacity, small *padding capacitors* (C1, C4, C7, Figure 2) are added to each tuned circuit. In this manner, the manufacturer can adjust the padder and set the minimum capacity of the tuned circuits to a specific value without fuss or bother. Since the receiving antenna has a slight detuning effect upon the first r-f circuit of the receiver, the padding capacitor for this stage (C1) is often made adjustable from the panel of the set, permitting the r-f stage to be peaked to match the antenna in use.

Once the minimum circuit capacity has been established, the balance of the circuit must be adjusted to tune to the desired low frequency as read at the opposite end of the dial. The low frequency limit reached by the circuit is determined by the size of the tuning capacitor and the inductance of the coil. Since it is not easy to alter the maximum capacity of the tuning capacitor without physically bending or changing the plates, the logical solution is to vary the inductance of the coil. This can be done by moving a powdered iron slug up and down within the coil. Thus the correct alignment of the r-f stage at the high frequency end of the band is adjusted by means of the padding capacitor shunted across the coil, and the alignment of the stage at the low frequency end of the band is accomplished by varying the slug within the coil.

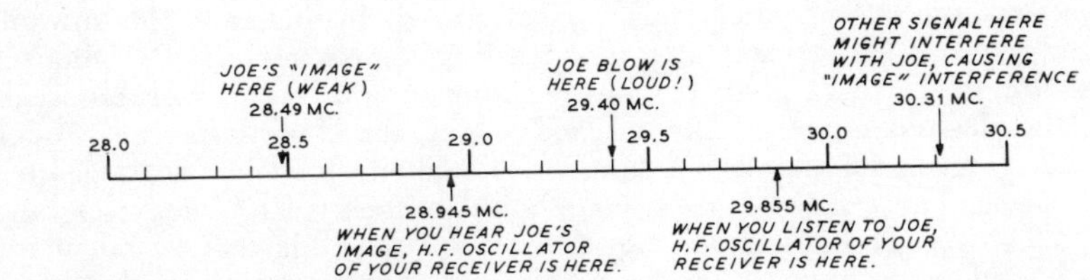

WHEN YOU LISTEN TO JOE:

REAL SIGNAL	H.F. OSCILLATOR (REAL SIG.+455 KC.)	YOU CAN HEAR "IMAGE SIGNAL (REAL SIG.+910 KC.)
29,400 KC.	29,855 KC.	30,310 KC.

WHEN YOU LISTEN TO JOE'S IMAGE:

YOU ARE LISTENING TO IMAGE AT:	H.F. OSCILLATOR (IMAGE + 455 KC.)	R.F. STAGES OF RECEIVER ARE TUNED TO:
28,490 KC.	28,945 KC.	29,400 KC.

Fig. 3 Image situation becomes confusing at the higher frequencies where both signals may be audible. Above chart shows relation between fundamental and image signal when receiver conversion oscillator operates on the "high side" of incoming signal. Conditions are reversed for "low side" operation of the oscillator. Most receivers employ "low side" oscillator operation on all but the highest frequency band, where "high side" operation is occasionally used.

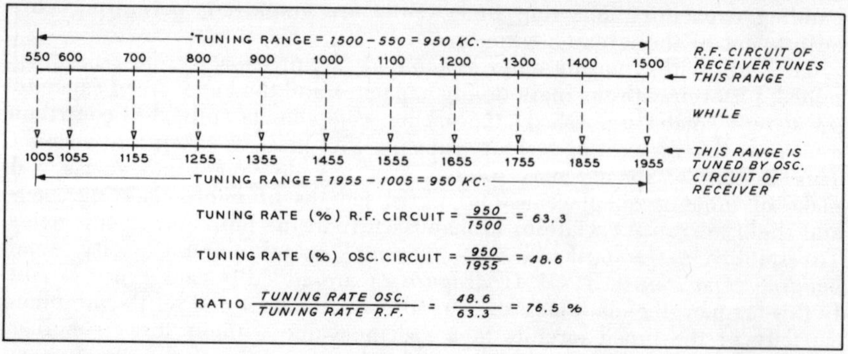

Fig. 4 Ratio of the tuning rate of h-f oscillator and r-f stages in superheterodyne receiver is different, although all stages tune through the same number of kilocycles. Shown above is relationship for broadcast band coverage of receiver employing 455 kc i-f amplifier. The tuning ratio is 76.5%.

Tracking the Tuned Circuits

As the receiver owner gaily tunes his receiver from one end of the dial to the other, he pays scant attention to the fact that the oscillator stage is racing along 455 kilocycles higher in frequency than that indicated on the receiver dial. When the receiver is tuned to 10,000 kilocycles, the oscillator is tuned to 10,455 kc. As discussed earlier in the receiver theory section, a superhet receiver tunes two frequencies at once, one higher in frequency than the conversion oscillator and one lower in frequency. The purpose of the r-f and mixer coils is to select the chosen frequency and reject the unwanted one. For ease of alignment, the signal frequency *lower* than the conversion oscillator frequency is normally used. In this case, the unwanted *image signal* should appear at a dial reading approximately 910 kilocycles *below* that of the real signal. The image signal should be considerably weaker than the desired signal *if* the r-f and mixer tuned circuits are properly aligned. For an example, let's assume local amateur Joe Blow is transmitting on 29.4 mc. The image of his signal should be heard 910 kilocycles lower, or at 28.49 mc. If Joe's image signal is louder than his real signal, it indicates one of two things: either the r-f circuits of the receiver are incorrectly tuned to the image signal, or the local oscillator of the receiver is operating on the "low" side of the signal. The receiver instruction manual will tell you which side of the signal is the correct place for the oscillator stage. Look at Figure 3. It summarizes this confusing situation and can help you keep things in order.

As you might guess, making the local oscillator maintain a respectful separation of 455 kilocycles from the received signal is "not all beer and skittles" (as they say in the old country). Some pretty fancy tricks must be employed in the receiver to maintain accurate alignment. Just consider the problem: To cover the broadcast band, the r-f and mixer stages must tune from 550 kilocycles to 1500 kilocycles. At the same time the oscillator stage must tune from 1005 kilocycles to 1955 kilocycles, always maintaining

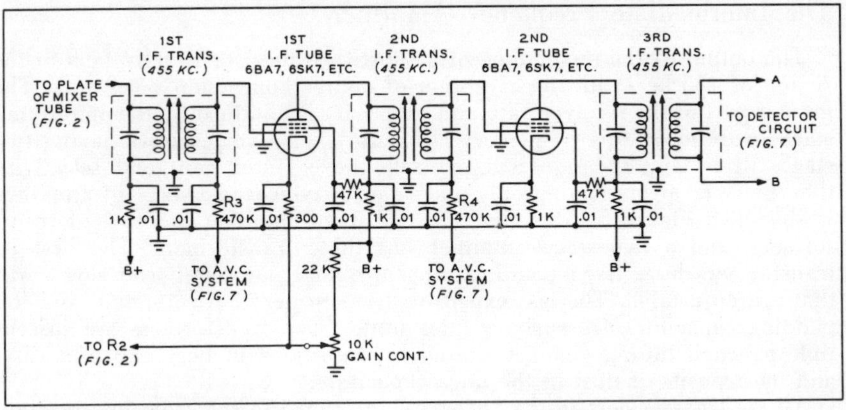

Fig. 5 Output signal from mixer tube is at a constant frequency of 455 kc, regardless of frequency of reception. The greater part of receiver gain and selectivity is obtained in the intermediate frequency amplifier, which is fixed-tuned to 455 kc signal of mixer tube. The i-f gain may reach 10,000.

the 455 kilocycle difference in frequency. This means that while the r-f and mixer stages are tuning *through a range of 950 kilocycles, the oscillator must tune through the same 950 kilocycle range.* (Watch out, now! Here comes the "joker"!) *The two tuning ranges expressed as a percentage of the highest frequency in each range are different!* The r-f and mixer percentage figure is 63.3, and the oscillator percentage figure is 48.8. Starting from the high frequency end of the dial, the r-f stages tune 63.3% lower in frequency, while at the same time the oscillator only tunes 48.8% lower in frequency. Therefore the *oscillator tuning rate* must be *slower* than that of the r-f stages because it covers a smaller percentage range. In this particular case, the *tuning rate* of the oscillator is only 76.5% as great as the tuning rate of the r-f and mixer stages. This is summarized in Figure 4.

The result of this complex situation is that the oscillator circuit must be modified in order to decrease the tuning rate. This may be accomplished in several ways. The most common way is to decrease the inductance of the oscillator coil, or the capacitance of the tuning capacity by the insertion of a *series padder* (See C10, Figure 1). Adjustment of the series padding capacitor will vary the rate of tuning of the oscillator.

If you will examine the usual run-of-the-mill ac-dc broadcast receiver you will note that no series padding capacitor is used in the oscillator stage. Rather, the variable plates of the oscillator tuning capacitor are cut to a different shape and size than those of the r-f section. This stunt can only be done in a receiver that covers a single tuning range, such as the broadcast band.

Summary: The high frequency oscillator must always be detuned to the high side of the signal frequency by an amount equal to the intermediate frequency. Since the percentage tuning rate of the oscillator is now less than the percentage tuning rate of the r-f circuits, a series capacitor must be added to the oscillator circuit to insure proper tracking between the stages.

The Intermediate Frequency Amplifier

The output of the mixer tube of a shortwave receiver usually falls in the region of 455 kc (with the exception of double conversion receivers). The greater part of the receiver gain and selectivity is obtained by the amplifying stages working at this frequency. The more expensive receivers having two stages of i-f amplification will naturally have better gain and selectivity than those receivers having only one stage. A typical two stage i-f amplifier is shown in Figure 5. Note that a single stage amplifier has two i-f transformers, and a two stage amplifier has three transformers. The best i-f transformers have fixed padding capacitors and powdered iron slug cores that are adjustable. The less expensive transformers have no slug, and the padding capacitors are variable mica units. Such transformers are susceptible to small tuning changes caused by variations in heat and humidity, and by deposits of dust in the mica capacitors.

All the transformers in the "i-f strip" are tuned to the same intermediate frequency of 455 kc. Each transformer has the property of accepting a narrow band of frequencies and rejecting all others. The better the transformers, and the more of them, the narrower is the passband of the receiver as shown in Figure 6. Selectivity, like all good things in life, cost money!

The Detector and A-V-C System

The next block of components to consider make up the detector and a-v-c system. Receiver manufacturers have as many different circuits to do these jobs as Carter has pills, and it is impossible to cover them all. However, they all have many points in common and a representative circuit is shown in Figure 7. The detector element is a simple diode, and may be one-half of a 6H6 or 6AL5, or perhaps the diode part of a 6SQ7 or 6AV6. The diode

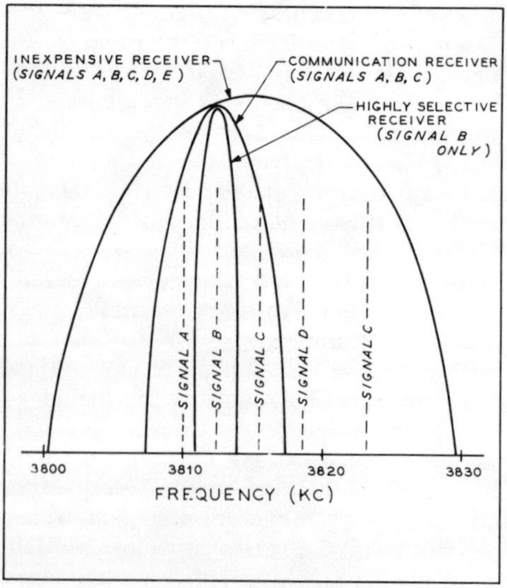

Fig. 6 Selectivity of the receiver is function of design of transformers in the intermediate frequency amplifier. To reduce inter-channel interference, good communication receiver must have passband less than 5 kc wide.

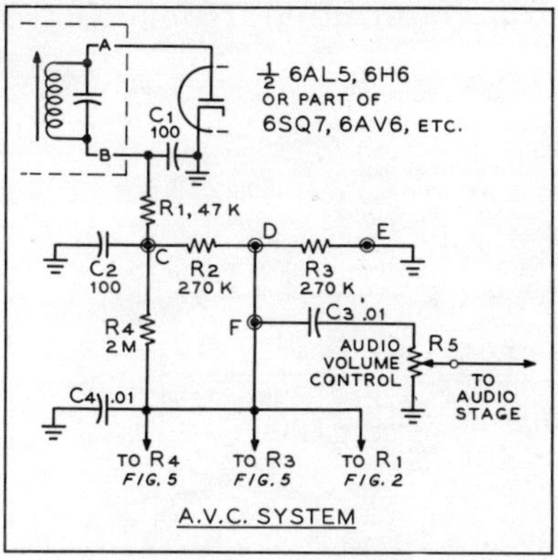

Fig. 7 Automatic volume control circuit reduces atmospheric effects on received signal, and prevents loud stations from "blasting" the receiver. Speaker volume is held constant by varying gain of the r-f and i-f stages in proportion to strength of received signal. Control voltage is obtained in the detector of receiver directly from the incoming radio signal.

rectifies the signal as explained earlier in this Handbook. The rectified signal is passed through an r-f filter (C1-R1-C2-R2) and through audio coupling capacitor C3 to the volume control, R5. A portion of the rectified signal is passed through a second filter (C4-R4) having a relatively long time constant which washes out the modulation on the signal, leaving a voltage that is proportional only to the carrier strength of the signal. This voltage is returned to the grid circuits of the r-f and i-f stages to control their gain. The greater amount of a-v-c voltage that is developed across C4-R4, the greater the amount of bias voltage that is applied to the a-v-c controlled tubes in the receiver. By a proper choice of circuit constants, the audio level of the receiver may be held constant for signal strength ratios of 10,000 to one.

The Noise Limiter

The purpose of a noise limiter in a shortwave receiver is to protect the eardrums of the listener from loud bursts of static, and also to reduce the racket produced by man-made sources. A simple diode limiter that removes the noise "peaks" from the signal is shown in Figure 8. Variations of this

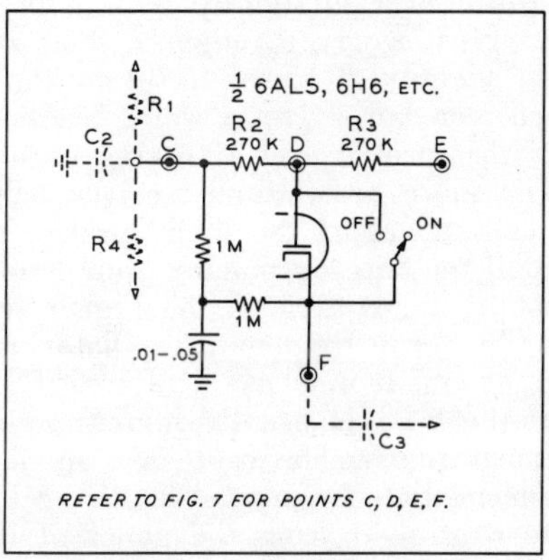

Fig. 8 Automatic noise limiter can easily be added to shortwave set to reduce static crashes and automotive ignition noise, which is particularly bothersome at the higher frequencies. "Popping" ignition noise is cut out of radio signal by action of diode tube. Five components and tube are all that are required to add noise limiter to most shortwave receivers.

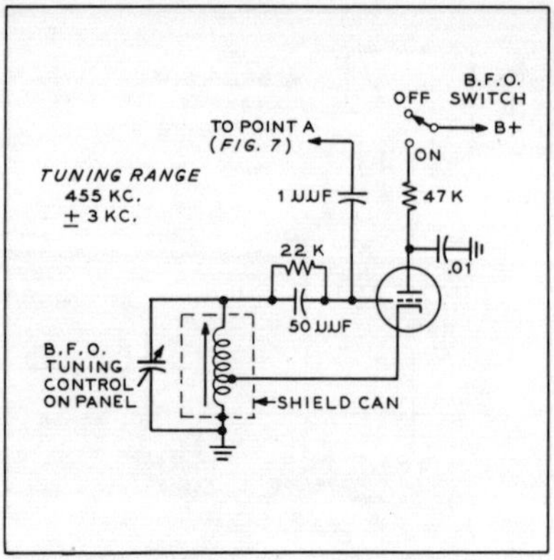

Fig. 9 Beat frequency oscillator is required for c-w and SSB reception. This circuit produces weak signal in i-f amplifier that is slightly detuned from incoming signal. The resulting "beat" or heterodyne makes code signal audible.

circuit are included in almost all receivers above the economy class. Operation of the device is automatic, since the bias for the noise rectifier is derived from the rectified signal appearing in the second detector circuit. When a loud noise pulse comes along the diode ceases to conduct, thus momentarily disconnecting the audio portion of the receiver from the detector. The simplicity of the device is apparent when it is seen that only four additional components need be added to the receiver to make the limiter! More complex limiters employing several tubes are employed in the larger receivers to provide better suppression of man-made interference.

Audio Stages, B-F-O, and Power Supply

The remainder of the receiver is very straightforward. The audio section may employ one or two pentode tubes in the output stage, along with a small audio amplifying stage. It is interesting to note that the less expensive receivers incorporate the speaker in the receiver cabinet, while the more elaborate receivers have the speaker contained in a separate box. It is a good idea to separate the receiver and speaker by a few feet, since the sound waves from the speaker can actually cause the wiring of the receiver to vibrate ever so slightly. This is of no concern on the broadcast band, and no great worry throughout most of the shortwave spectrum. However, in the vicinity of 15 mc to 30 mc the minute vibration of the r-f leads in the receiver can set up a weird feedback effect that will produce a loud howl in the speaker and a ringing noise that is most annoying and mystifying. The more pretentious receiver helps to solve this acoustical problem by the removal of the loud speaker from the receiver cabinet. Please don't spoil the cure by placing your loudspeaker atop the receiver! You are just asking for trouble at the higher frequencies!

The beat frequency oscillator is a standard adjunct to all shortwave receivers. It may use a 6C4, 6SJ7, 6J5, or 6AU6 tube. The coil assembly of the b-f-o is usually housed in a can similar to those encasing the i-f transformers. The b-f-o is a small oscillator that operates at or near the intermediate frequency of the receiver. The frequency of the oscillator may be controlled from the panel of the receiver. When the b-f-o is slightly

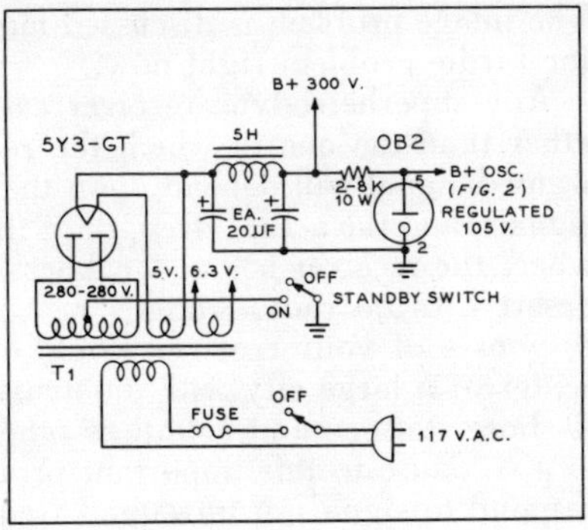

Fig. 10 Transformer-type power supply is standard equipment in communications receiver. High frequency oscillator voltage may be regulated by gas-filled control tube to achieve greatest frequency stability and immunity to line voltage fluctuations. Supply is fused for safety.

detuned from the intermediate frequency, it will create a beat note or hetrodyne with the signal in the i-f circuit. This is quite a nuisance for phone reception, but a tremendous help when receiving c-w signals and single-sideband stations. A typical triode b-f-o circuit is shown in Figure 9. In passing, it should be noted that the a-v-c system of the receiver must be disabled when the b-f-o is in use, or else the b-f-o signal would control the a-v-c voltage and desensitize the receiver. Most receivers couple the b-f-o and a-v-c switch together so that when the b-f-o is turned on the a-v-c system is shorted to ground, as shown in the schematic.

The power supply of the average short wave receiver is a rather unimaginative device. Ninety-nine times out of one hundred it consists of a capacitor input filter system (Figure 10). The better receivers incorporate an additional voltage regulator tube which is used to insure a stable d-c plate voltage applied to the high frequency oscillator of the receiver. A 5Y3 rectifier is almost always used in a supply of this type.

Miscellaneous Receiver Controls

A few other knobs decorate the panel of the shortwave receiver. The B-plus voltage may be interrupted by a *standby switch,* rendering the receiver inoperative during periods of transmission of a nearby transmitter. A potentiometer in the cathode circuit of the r-f tube may be used to reduce the r-f gain of the receiver for c-w and single-sideband reception.

Spurious Responses

The true criteria of a good receiver is *not* what it receives, but what it *rejects!* You can hear a lot of DX on most any old "clunker" of a receiver, but you will also hear a lot of spurious signals that are not supposed to be there! Since it is prohibitively expensive to design a tuned circuit that will pass only *one* frequency, stations on either side of the wanted one will usually be heard, and will cause interference. In addition, image response and "birdies" will place signals on your dial that you wish weren't there!

The image problem is discussed later in the Handbook, and we can examine the birdie problem right now.

Any superhetrodyne receiver can have spurious responses at frequencies other than the one to which the receiver is tuned. The response to the unwanted signal will depend upon the susceptibility of the receiver to spurious signals, and the actual strength of the radio signal that happens to be situated where the receiver has a "weakness." If you are in the middle of the Sahara desert with no radio stations within a thousand miles of you, the spurious responses of your receiver would cause you little worry. If you are in the center of a large city with amateurs, broadcast stations, police radios, ship-to-shore stations and countless others on all sides of you, it would be wise to give pause to this important problem. If your receiver has a tendency to respond to signals at unwanted frequencies you might hear the local broadcasting station in the middle of the 80 meter ham band, or the nearby police radio station squarely on top of the BBC in the 31 meter band. If you have never heard a birdie, you might just place a good, long antenna (200 feet or so) on your receiver and listen to the 160 meter region. Chances are—if you are located near a big city—you will hear a jumble of broadcast stations that are not really there. The spurious responses of your receiver magically produces them. If you shorten your antenna and decrease the signal pickup, the spurious responses will drop in strength and may even vanish.

There's not much you can do to the receiver to drop the level of spurious susceptibility. It's a sort of a built-in factor, and you are stuck with it. Some of the newer r-f tubes, such as the 6DC6 and the 6BZ6 are especially designed to cope with this problem. About the best thing to do is to place additional selectivity *ahead* of the receiver to reduce the level of the unwanted signals reaching the r-f circuits of the receiver. The preselector described in a later chapter should help greatly to eliminate spurious responses that may bedevil you.

Selectivity Aids for the Communication Receiver

Three items are in use today that help to obtain optimum selectivity in the communications receiver. They are the crystal filter, the Q-multiplier, and the mechanical filter. In each case, the idea is to narrow the bandwidth of the receiver, permitting the reception of only one station at a time. This function is accomplished in the fixed-tuned intermediate frequency amplifier stages.

The Crystal Filter

The crystal filter is a sharply tuned stage employing a resonant circuit made of a piece of critically ground quartz. The quartz crystal is tuned to the frequency of the i-f amplifier, and acts as a signal "gate," permitting the intermediate frequency to pass, and rejecting those frequencies on either side. The crystal filter is extremely effective for c-w reception, but does not perform nearly as well for phone reception. The crystal filter allows a high degree of rejection at *one* chosen frequency. By the use of a variable rejection control, a loud unwanted hetrodyne may be almost eliminated.

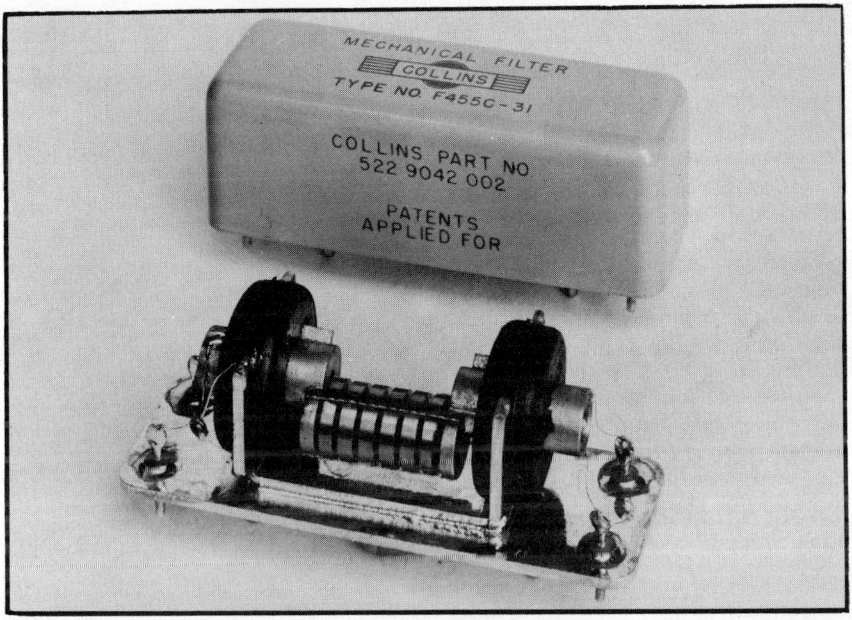

COLLINS MECHANICAL FILTER offers ultimate in receiver selectivity. This unit is a selective electro-mechanical device designed to operate in the intermediate frequency circuit of the receiver. It provides a passband having very high rejection properties to adjacent signals, as illustrated in Figure 6. The bandwidth of the mechanical filter is fixed, and different filters must be switched in and out of the circuit to permit various degrees of selectivity to be used. Collins 75A-4 receiver incorporates filter switch having three positions for 6 kc, 2.8 kc, and 500 cycle bandwith filters. The 500 cycle filter is used for c-w operation.

The Q-Multiplier

The Q-multiplier is a relatively new device for increasing the selectivity of a receiver. Electronically speaking, this circuit increases the "Q" (or figure of merit) of a coil in the i-f amplifier to the point where the tuned circuit employing this coil will pass only a narrow band of frequencies. This is by far the simplest and most inexpensive means that the Novice can employ to increase the selectivity of his receiver, therefore a Q-multiplier is described in detail in the chapter devoted to Novice aids. The Q-multiplier may be used for either phone or c-w reception.

The Mechanical Filter

The Collins mechanical filter employs small resonant metal discs that are tuned to the i-f frequency of the receiver. The passband of the mechanical filter is excellent, and it is by far the best (and most expensive) device for obtaining maximum receiver selectivity. The passband of this type of filter is fixed, and various models of the filter allow the use of different passbands for phone or c-w reception. Its characteristic broad flat-top and steep sided response curve make it a pleasure to copy phone or c-w even under the worst QRM conditions.

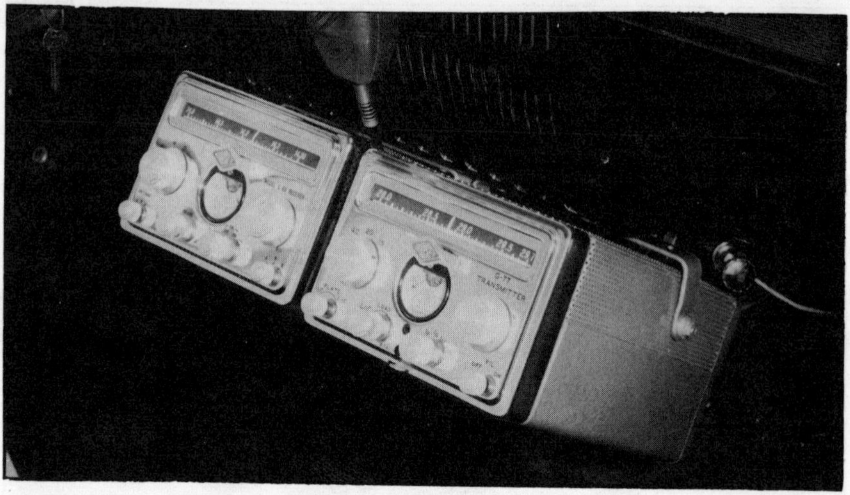

MOBILE TRANSMITTER AND RECEIVER, such as the Gonset "Twins" (above) allow all-band operation of amateur radio station while automobile is in motion.

CONVERTERS

There is one form of "receiver" that has not yet been discussed, and that is the *converter*. A converter is a device that will extend the tuning range of a receiver. Take your home radio for example: If you desired to listen to short wave signals, you could build a converter similar to those shown in this Handbook which would extend the tuning range of the b-c set up into the short wave bands. Admittedly, the combination would not make a very good communications receiver, but it would work.

Converters are divided into two groups, tunable and fixed-tuned types. Both have decided advantages and disadvantages that make one type preferable to the other for certain applications.

Either type of converter may be used in conjunction with an inexpensive shortwave receiver to improve reception at the higher frequencies. As an example, most of the inexpensive communications receivers will not perform to a high degree on the 10 meter band (28-29.7 mc). It is possible to build a converter that will "beat" the 10 meter signals down to a lower frequency at which the receiver will be more sensitive. Converters are also used to an advantage in conjunction with an automobile receiver to facilitate mobile operation of an amateur station.

The converter, in effect, is comparable to the "front end" of a communication receiver. The receiver to which the converter is attached serves as the i-f amplifier and detector stages for the r-f circuits of the converter.

A block diagram of a typical fixed-tuned converter is shown in Figure 11 The r-f amplifier section is adjusted so that it will amplify a band of frequencies between 50 and 54 mc. The conversion oscillator is crystal controlled at 43 mc. Note that the conversion frequency is fixed, and cannot be changed unless a new crystal is inserted in the oscillator. Any signals

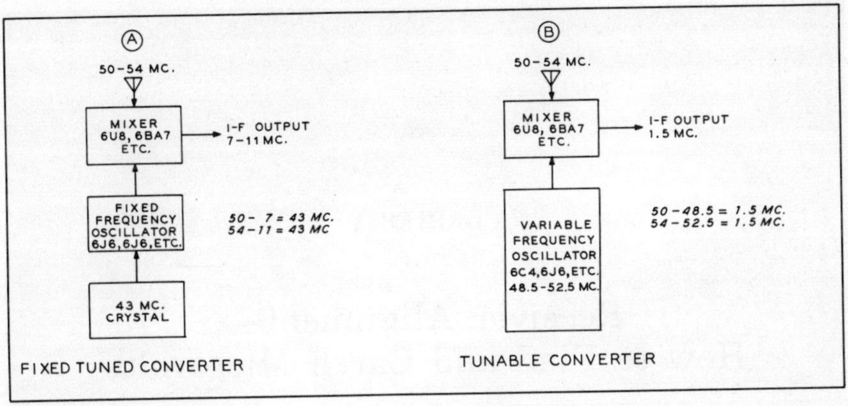

in the 50-54 mc range are mixed with the 43 mc oscillator signal in the 6U8 mixer tube. The i-f frequency range for reception across the six meter band is the difference in frequency between the tuning range and the conversion oscillator frequency. In this case, the i-f range is 7 mc to 11 mc. If the converter is attached to a communication receiver capable of tuning this frequency range, signals in the 50-54 mc range may be received. Note that the r-f stage of the receiver serves as a tunable i-f stage for the converter. The converter-receiver combination actually has two i-f channels, and is called a *double conversion receiver.* With a fixed-tuned converter, all tuning is done with the main tuning dial of the receiver. The converter can be tucked out of the way and left un-attended for long periods of time. Another advantage is that the crystal controlled conversion oscillator in the converter contributes negligible drift to the 50 mc signals. On the debit side of the ledger, this type of converter has lower gain than the tunable type, because the r-f stages must be adjusted to cover the complete band of frequencies to be received. The broad-band r-f stage will sometimes let signals get through it which are unwanted. These are known as *spurious signals.* In spite of this drawback, this type of converter is unanimously favored over the tunable converter for high frequency reception.

Figure 11 shows a typical tunable converter diagram. The oscillator section of this converter, unlike the crystal controlled model, can be changed in frequency by means of the main tuning dial. When used with an automobile radio, the i-f channel of the converter is 1500 kc. The car radio is tuned to that frequency, and the conversion oscillator of the converter always operates 1500 kc away from the received signal. A fixed intermediate frequency is used, and all tuning is accomplished at the converter. This type of converter features good gain, good selectivity, and good spurious signal rejection because all the r-f tuned circuits are sharply resonant at the incoming signal frequency. The disadvantages of this type of converter are the large size (because of the bulky tuning capacitor), thermal drift (because of the heating of the oscillator components) and relatively poor mechanical stability, especially in an automobile. In most cases, it is usually not as handy to tune the converter as it would be to tune the receiver. However, this type of converter is by far the most popular type for mobile reception.

CHAPTER V

Receiver Alignment—
How to Find and Catch Microvolts

Have you ever lifted the lid of your receiver and poked around in the insides? Have you removed the bottom dust cover and adjusted the trimmers to improve reception? Have you tested the tubes to see if they are below par? If you have, you are a proud member of a small minority of listeners. Most hams and SWL's shy away from the "innards" of their receiver as if there was a flame-thrower within, waiting to roast them to a crisp.

This chapter takes you on the hunt of the elusive microvolt, and shows you how it may be captured and tamed for the benefit of better shortwave reception. Put on your pith helmet, take your alignment tool in hand, and away we go!

Inside your receiver are a multitude of small adjustment points called *trimmers.* These vernier adjustments control the sensitivity, selectivity, image rejection, and dial calibration of your set. If these adjustments are in perfect alignment, the receiver will perform to the best of its ability. When they are "out of whack" the receiver is subject to image interference, and it may become insensitive to all but the loudest signals, or the dial calibration may go awry.

The purpose of the trimmers is to insure that all the r-f circuits track perfectly as you tune the dial of your receiver. They compensate for slight variations inherent in the components and tubes of the receiver, and permit the tuned circuits to be placed "on the nose" before the receiver is sold.

Why worry about these adjustments? "If my receiver is new, haven't all these little jiggers and plungers been properly set by the manufacturer?", you ask. Well, perhaps not. Remember that your receiver is built and aligned on a production basis. The alignment is not done by an engineer, but rather by a technician trained for the job, who may or may not know anything about radio receivers in general, or shortwave reception! Perhaps the technician had a fight with his wife the night before, or perhaps came to work with a bad headache, or is worrying about the race for the pennant in the National League. Is he going to worry about the *precise* alignment of the particular receiver that you might chance to buy? Don't be silly. He will do

Fig. 1 Placement of major components of National NC-66 all-wave receiver can be seen in top view. Bandspread tuning capacitor is to left of main tuning gang. The intermediate frequency amplifier runs across the back of chassis, with the speaker output transformer at right. Broadcast band ferrite-type antenna coil is mounted above the chassis to the rear. The speaker is below chassis, at right.

his best, and if the alignment job passes the minimum specifications demanded by the manufacturer, he will be satisfied. After the receiver leaves the factory it is trundled about in a box and jolted around at the mercy of the shipping crew and the not-so-tender hands of the freight carrier. The many uninterested hands pass the set along until it arrives at your house, ready to play. You take your chance on alignment accuracy, and hope for the best!

Let's look again at your receiver after you have had it for a year or so. The tubes and parts have aged and their electrical characteristics have changed a bit. Minute dust and dirt particles have sifted into the coils, capacitors, and switches of the tuned circuits. As a result, the important selectivity and sensitivity adjustments have slowly deteriorated over the months—so slowly that you have not noted the change. But your DX-bloodhound has gradually turned into a mongrel, and you are listening to all the stations through a thick veil of mis-alignment. You won't notice this gradual deterioration on a broadcast receiver tuned to a 50 kilowatt station a few miles away, but the change is deadly in the shortwave spectrum where the receiver is dealing with signals of microvolt intensity.

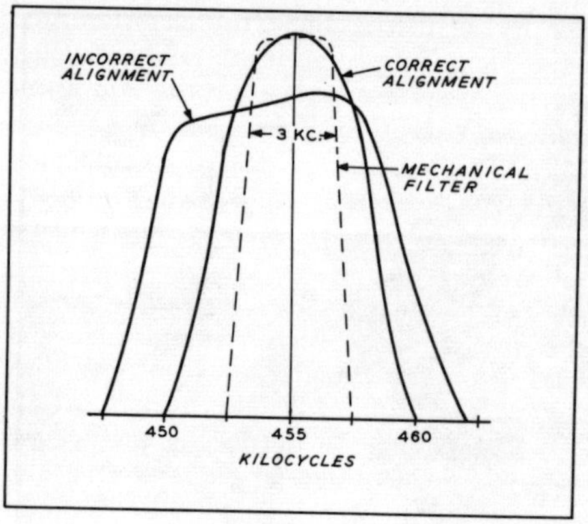

Fig. 2 Selectivity and gain of the receiver are determined by the i-f amplifier. The correctly aligned amplifier has symmetrical curve with rounded shoulders. A poor alignment job will result in lop-sided curve with tilted top. Ideal response curve is shown by dotted line. Mechanical filter produces response curve with relatively flat top and steep sides.

Let's Tweak An Adjustment—Just For Fun

Open the instruction book of your receiver and find the *Service and Alignment section.* Read it through and study the diagrams and alignment information. Particularly note the section dealing with alignment of the intermediate frequency (i-f) amplifier. Open the lid of your receiver and find the i-f transformers and tubes. Note that each transformer has one or two trimmers on it. The trimmers may be located on the top or the side of the transformers. If only one trimmer is on the top, the chances are that the second one is on the under-side of the transformer, and may be adjusted from the bottom of the set. These trimmers should be adjusted with the aid of an insulated screwdriver for two very important reasons. First, the trimmers may be "hot" with high voltage. If you touch them with the metal blade of a screwdriver you *might* get an unnerving and nasty shock. Second, the metal of the screwdriver might detune the circuit, causing a maladjustment in tuning. Many varieties of alignment tools are available in the large radio stores for less than a dollar. The *ICA #6192* alignment tool should do the trick for you.

The layout of a typical i-f amplifier is shown in Figure 1. Each stage is tuned by means of compression-type mica capacitors placed across each winding of the transformers, or by adjustable powdered iron slugs placed within the windings. The heads of these trimmers may be easily turned with the blade of the alignment tool with no danger of shock or misalignment.

Regardless of the frequency to which the receiver is tuned, the conversion process of the mixer stage changes the frequency of the signal down to the intermediate frequency, which is usually in the neighborhood of 455 kilocycles. This frequency is used, since it is possible to achieve more gain and selectivity per stage at a low frequency than at any frequency in the broadcast or shortwave spectrum. This is why almost all intermediate frequency amplifiers of modern receivers are tuned to some spot in the 200 kc to 500 kc frequency range. Selectivity and gain go hand-in-hand with inexpensive components and ease of adjustment. All i-f transformers can be tuned over a very narrow range, and they should be tuned to the single intermediate frequency of the receiver by adjusting the trimmers mounted on the transformers. If one transformer of an amplifier is tuned to 450 kc, and another one is tuned to 457 kc, the response curve of the amplifier will look like curve B of Figure 2. Sad, isn't it?

A combination of pentode tubes and high gain i-f transformers may provide a signal gain as great as 10,000. In addition, the selective circuits of the i-f transformers reject signals off the resonant frequency, contributing a major portion of the selectivity of the receiver.

All right now—have you found the adjustments on the i-f transformers? Tune in a local broadcast station and shorten the receiver antenna so that the station provides a mid-scale deflection on the S-meter of your set. Pick out one of the i-f transformer trimmers at random, and make a mental note of the position. Pencil a mark on the transformer case indicating the setting of the trimmer, if you wish. Now, take your insulated alignment tool and slowly turn the trimmer one way and then the other. Watch the S-meter while you do this. You will note that the adjustment is not critical, and the signal will get a little weaker, or a little stronger as you move the trimmer. Adjust the trimmer for maximum reading of the S-meter. Check and see if the trimmer is in the same position as when you started the test. If not, this adjustment was out of alignment, and you've corrected the error.

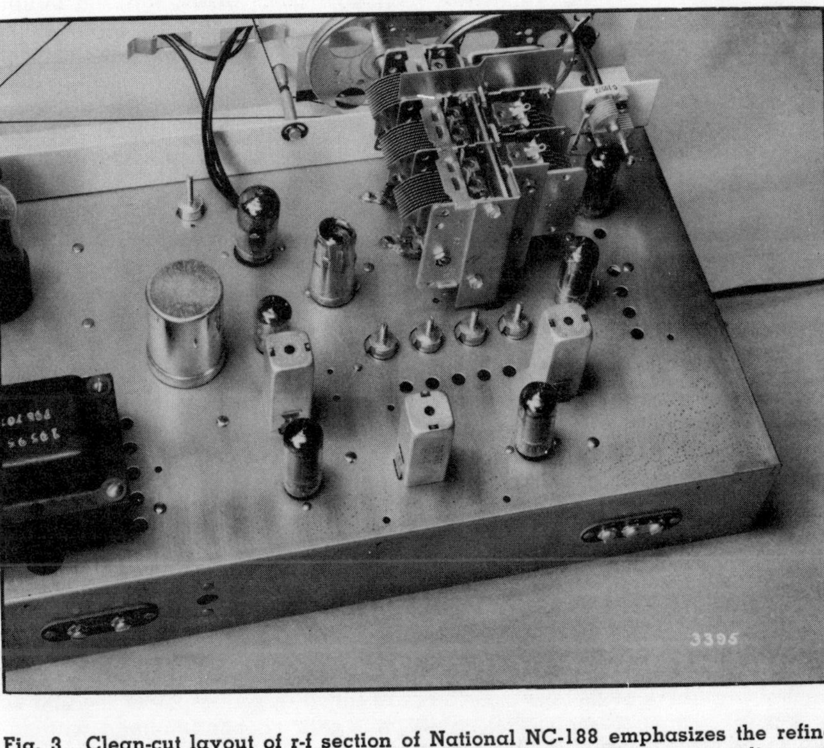

Fig. 3 Clean-cut layout of r-f section of National NC-188 emphasizes the refinement of design. The oscillator tuning slugs are behind the tuning capacitors and i-f transformers (Z1, Z2, Z3) and associated tubes are at the rear of the chassis.

I-F Amplifier Alignment

Now that you have the "hang" of it, you can align the whole i-f amplifier. Start with the *last* i-f transformer (the one between the last i-f tube and the second detector) and progress forward towards the mixer transformer. If your receiver has a crystal filter, you had better leave the adjustments on it alone for the time being, as it is a foxy business to properly align the filter. It is necessary, however, to make sure that the i-f amplifier of your receiver is properly aligned to the crystal frequency of the filter. Be sure you tune your receiver so that the test signal is properly centered in the pass-band of the crystal. To accomplish this, switch on the filter and tune the receiver for highest S-meter reading of the test signal. Take it easy! The crystal peak is very sharp. When you are "on the nose," you can align your i-f transformers for maximum reading of the meter. Check the receiver tuning. If the receiver drifts a bit, you might have to repeak the signal to prevent it from drifting off the peak of the crystal. With practice, you can actually align the i-f stages by ear, and you'll probably have to do this if your receiver has no S-meter.

As an example, a top view of the chassis of the *National NC-188* receiver is shown in Figure 3. I-f transformers Z1, Z2, and Z3 are marked. The iron core trimmers may be adjusted through holes in the top and bottom

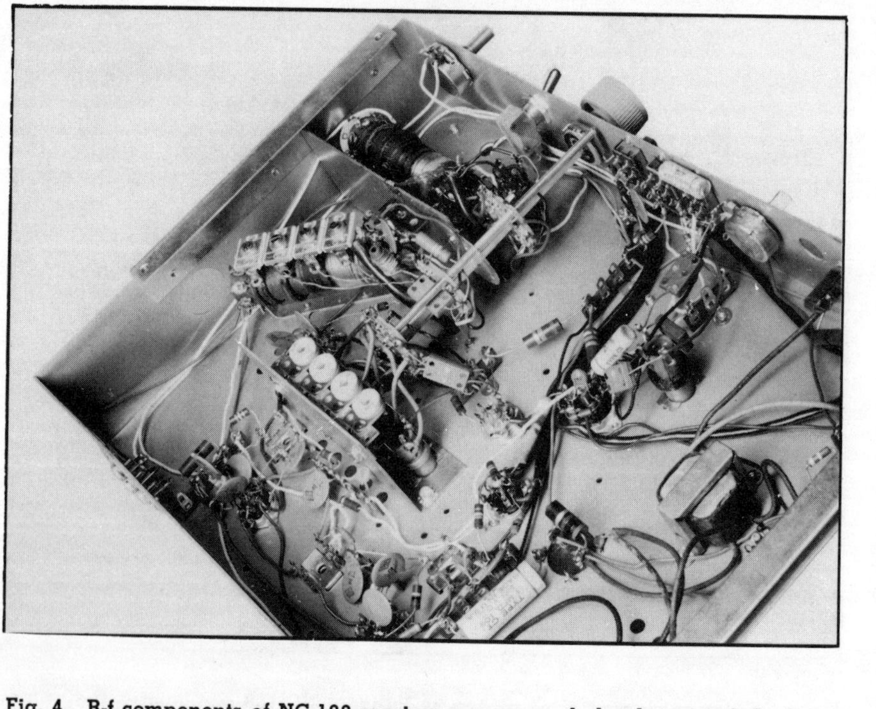

Fig. 4 R-f components of NC-188 receiver are grouped closely around the bandswitch. R-f coils are at upper left, with mixer coils directly below. The oscillator coils and the trimming capacitors are directly behind the bandswitch.

of the transformers. The bottom slug holes can be seen in Figure 4. Start the alignment from the "rear end" of the i-f amplifier. Align the slugs of Z3 first, then Z2, then Z1. Tune the slugs for maximum S-meter reading of a moderately strong signal. It makes no difference which transformer slug you adjust first, and you can change the order of alignment if you wish, but the "tail-end-first" method is generally used for a score of obscure reasons.

While you are at it, you might as well align the beat frequency oscillator (b-f-o). In most receivers, a triode tube is used as the b-f-o for c-w and single sideband reception. The purpose is to provide a heterodyne against the incoming signal so that code or SSB transmission can be made intelligible. The b-f-o is actually a miniature signal generator, tuned to the intermediate frequency and coupled to the i-f amplifier. If the b-f-o is tuned to 456 kc it will produce a one kc audible beat with an i-f signal on 455 kc. Because the a-v-c circuit of the receiver will rectify the signal of the beat oscillator, producing a strong a-v-c voltage and a consequent drop in receiver gain whenever the b-f-o and a-v-c are both functioning, the a-v-c is normally disconnected whenever the b-f-o is used. For purposes of b-f-o alignment, however, the a-v-c may be left on, if desired. The antenna is removed from the receiver and the b-f-o tuning control *(pitch* control) on the panel is set at the center position. This permits the control to rotate an equal number of degrees to the right and to the left. With the knob in the center, the b-f-o should be tuned to the intermediate frequency of the receiver. Find the adjustment control on the beat oscillator transformer. The trimmer of the NC-188 b-f-o is shown in Figure 3. It is the slug-tuned coil (L12) mounted behind the front panel. Turn on the b-f-o and note the hiss heard in the speaker. Tune the trimmer of the b-f-o transformer back and forth, and adjust it so the hiss has the lowest pitch. Turning the trimmer in either direction from the correct setting should raise the pitch of the hiss. If the trimmer is turned far enough in either direction, this hiss will rise in pitch and gradually get weaker. Bring it back into audibility and adjust the trimmer as described. The beat oscillator is now tuned to the center of the i-f passband. It may be detuned slightly to one side or the other of the passband by varying the vernier panel control. In many instances, this pitch control is a variable capacitor, but in the NC-188, it is a potentiometer (R32), seen in Figure 4.

Your i-f stages are now properly aligned. It is well to check the alignment every year or so to make sure that the adjustments do not drift out of alignment. In most cases the stages will "stay put," but it is wise to make sure. Also check alignment whenever the i-f tubes are changed. *Summary: The i-f circuits of your receiver are aligned by trimmer adjustments mounted within the i-f transformers. These trimmers control the gain and the selectivity of your receiver. Adjust the trimmers for maximum strength of a steady, local signal. Peak each trimmer for maximum S-meter reading of the signal.*

R-F Alignment Of Your Receiver

The main drawback of the superhetrodyne receiver is that it is capable of simultaneous reception on two different frequencies at the same time.

The unwanted frequency is called the *image frequency*. Reception of the image signal may be substantially eliminated by the use of a radio frequency amplifier stage placed ahead of the converter stage of the receiver. The image frequency is displaced from the wanted frequency by twice the amount of the intermediate frequency. Thus, if we were listening to a station on 14,000 kc with a receiver having a 455 kc i-f stage, a signal on the image frequency of 14,910 could be heard in the receiver. A good r-f stage can amplify the 14,000 kc signal and reject the image signal. Many of the inexpensive short wave receivers on the market have no r-f stage for reasons of economy, and these receivers are bothered by unwanted image signals. The r-f amplifier stage also increases the overall sensitivity of the receiver to a great degree. No shortwave receiver can be called a true communications receiver unless it has a good r-f amplifier stage ahead of the first converter tube, similar to that shown in Figure 5

R-f alignment, therefore, is extremely important since maximum sensitivity and optimum image rejection both depend upon accurate tuning of the r-f and mixer circuits of the receiver. Look at the bottom view of the NC-188 receiver shown in Figure 4. The r-f stage coils are located in the far corner of the receiver. The antenna and grid coils for the three lowest frequency bands (L1A, L1B, L1C) are wound on one form, and the r-f coil for the highest frequency band (L2, 14-40 mc) is wound on a separate form, mounted next to the bandswitch deck, S1A. The schematic of this section of the receiver is shown in Figure 6. This receiver has an auxiliary antenna trimmer (C1) mounted atop the chassis, adjacent to the 6BA6 r-f amplifier tube. All that is required to bring the circuit into alignment is to tune the trimmer for maximum strength of the incoming signal. Easy, isn't it? There are no other adjustments that need to be made to this group of r-f coils.

Other makes of receivers, however, may employ separate r-f trimmers connected across each coil, as in the case of the *Hallicrafters S-85* (Figure 7). Each high frequency r-f coil has its own trimmer capacitor, in addition to having an adjustable slug for the high frequency coils. No panel antenna trimmer control is required. The individual bands are adjusted for maximum sensitivity by changing the inductance of the coil (L1, L2) at the *low frequency* end of the tuning range, and adjusting the trimmer capacitor (C1, C2) at the *high frequency* end of the tuning range. The trimmers that accomplish this are shown in the under-chassis drawing of Figure 8.

Assume the "number three' band (4.5—13.0 mc) is to be aligned. The

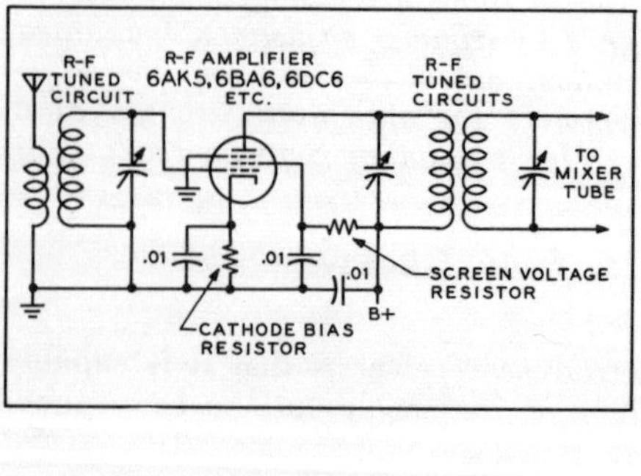

Fig. 5 Simple radio frequency amplifier placed ahead of mixer tube reduces tendency of superheterodyne circuit to receive two frequencies at once. R-f amplifier reduces the response to unwanted signals.

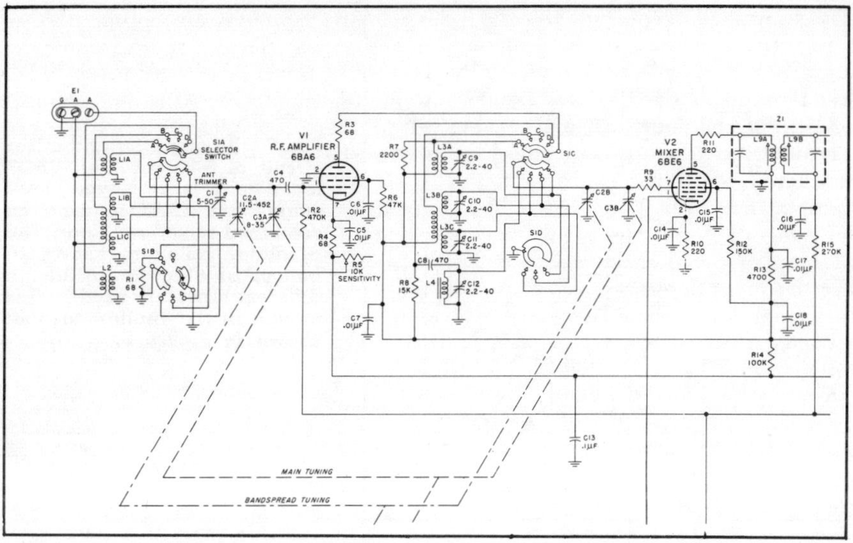

Fig. 6 R-f circuit of National NC-188, showing coil selector switch and bandspread tuning arrangement. Low noise 6BA6 miniature pentode is used in r-f amplifier for optimum signal-to-noise ratio, and good a-v-c response.

receiver is first tuned to a steady signal near 5.0 mc. This signal might be an r-f generator attached to the antenna terminals of the receiver, or it might be a loud station operating near that frequency. The slug of antenna coil L2 is tuned for maximum signal as read on the S-meter of the receiver. When the slug has been adjusted properly, the receiver is tuned to 12 mc, and the capacitor trimmer (C2) is adjusted for best signal response at this frequency. These adjustments mutually affect each other, and tend to be slightly interlocking. If much change in the trimmer setting is made at one end of the band, the other adjustment probably should be retouched for maximum response.

The same procedure is followed for the mixer stage of the receiver. If the mixer coils have tuning slugs, they should be adjusted for maximum signal response at the low frequency end of the tuning dial. The trimmer capacitors are adjusted for maximum response at the high frequency end of the dial. You should adjust the coil slugs first, then follow up by adjusting the trimmer capacitors.

The mixer circuit trimmer capacitors of the NC-188 receiver may be seen in Figure 4 (C9, C10, C11, C12) and in the circuit of Figure 6. The mixer coils for the three low frequency bands are adjusted at the factory and do not require slug manipulation. The mixer coil for the high frequency (D) band is slug tuned by a small powdered iron core threaded inside of the form. Adjustment of the slug is made near the low frequency end of the dial.

In most cases—especially if the receiver is new and does not exhibit any "dead spots" in the tuning range—the adjustment slugs can be left alone, and all minor alignments can be made at the high frequency end of each

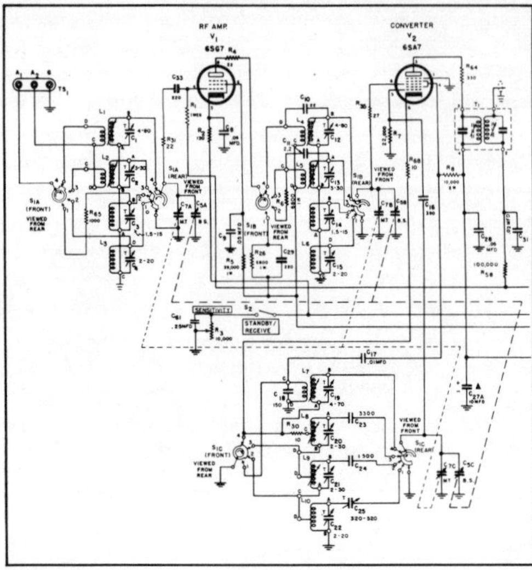

Fig. 7 R-f amplifier and mixer circuit of S-85 receiver. 6SG7 metal tube is employed in r-f stage, and 6SA7 is used as combined mixer-oscillator. The oscillator coils are shown at bottom of diagram. Although different tubes are used in this circuit, it is similar to that shown in Figure 6.

dial range by peaking the trimmer capacitors. If you have no signal generator, this can be done by listening to the background noise level of the receiver. Adjust the trimmer for maximum background noise at the high frequency end of each band. If the background noise drops off toward the low frequency end of the tuning dial, it indicates the coil adjustment must be varied a bit to "pep" up reception at that end of the dial.

The mixer circuit of the S-85 is shown in Figure 7. The two high frequency coils (L4, L5) are slug tuned for low frequency adjustment, and have trimmer capacitors (C12, C13) for high frequency alignment. The two low frequency coils are adjusted at the factory for proper inductance, and have capacitor trimmers (C14, C15) for alignment at the high frequency ends of the respective bands. Compare the circuits of Figure 6 and Figure 7 with the "front end" circuit of your receiver, and you can easily spot the adjustment points in your set. You can approach these adjustments with impunity, as none of them will affect the dial calibration of the receiver. The r-f and mixer circuits can actually be peaked by ear, or with the aid of the S-meter and a few steady signals. The time spent in checking the alignment of these circuits will pay big dividends in better reception.

Recently the author bought a new all-wave receiver in the $300 price class. It performed well, but seemed to be a little "flat" on the 10 and 15 meter amateur bands. Signals could be heard, to be sure, but the set didn't exhibit the "sock" that should have been apparent. After a little examination, it was noticed that the 10 meter image signals were almost as loud as the fundamental signals. Peaking the r-f trimmer control on the panel of the receiver made little difference in the degree of image rejection or receiver sensitivity. Acting on the hunch that the mixer stage circuit might be misaligned, the receiver was removed from the cabinet and turned on end. A quick perusal of the instruction book soon located the trimmer capacitors of the mixer stage. The receiver was tuned to the high frequency end of the dial (31 mc) and the mixer trimmer was carefully adjusted for maximum background noise. The receiver awoke with a bang, as the background level picked up with just a small tweak of the trimmer. Ten meter signals appeared as if from nowhere when the receiver was tuned across the ham band. The image signals were suitably attenuated. When the r-f trimmer on the panel of the set was tuned through resonance the peak in background noise could clearly be noticed. Within five minutes of the time the receiver was taken from the cabinet, the adjustment was made, and the receiver was returned to the cabinet in good working order. *Summary: The r-f and mixer tuned circuits of the receiver determine the sensitivity and image rejection of the receiver. To insure proper tracking of the circuits across a whole band, the inductance of the r-f and mixer coils in use must be adjusted for maximum signal response at the low frequency end of the tuning range. The trimmer capacitor for each stage must then be adjusted for maximum signal response at the high frequency end of each range. Since the inductance of the coils is not subject to drastic changes as is the capacitance of the trimmers, it may often suffice to peak the circuit padders for optimum reception near the high frequency end of each tuning range.*

The High Frequency Oscillator

The *converter stage* of the receiver (often called a *first detector* or *mixer)* changes the frequency of the incoming signal to the frequency of the fixed-tuned intermediate frequency amplifier. A special multi-grid converter tube may be used, one portion acting as the mixing unit, and the second portion serving as the high frequency (conversion) oscillator. These functions may also be accomplished by two separate tubes. The small r-f signal generated by the oscillator is combined with the incoming signal within the mixing tube, and the resulting "beat" signal is generated at the intermediate frequency, ready to fall into the waiting arms of the i-f amplifier.

As the dial of the superheterodyne receiver is tuned, the high frequency oscillator follows the frequency of reception, but always at a fixed distance away from it in frequency. This frequency difference is arithmetically equal to the intermediate frequency, as discussed in chapter 4. Since the mixer can act upon all signals that are removed from the desired frequency by the amount of the intermediate frequency, it is the job of the r-f stage to pick out the desired signal and to reject the undesired (image) signal. So far, so good. The main goal in life of the high frequency oscillator, therefore, is to tag along a respectful distance away from the r-f circuits and to supply the necessary beating signal at this frequency.

Yes, it is a genuine radio signal, too! The high frequency oscillator of any receiver is actually a pee-wee radio transmitter, generating a flea-power signal that may be heard some distance away from the receiver. During the last war this interesting fact was put to good use by the U.S. Navy. They would search for and find the frequency the enemy used to send messages to his naval ships. Since the ships maintained radio silence, and only used their receivers, the enemy felt sure that the ships could not be detected. The Navy found the frequency of transmission to the ships, and then tuned sensitive direction finders to the frequency of the conversion oscillator in

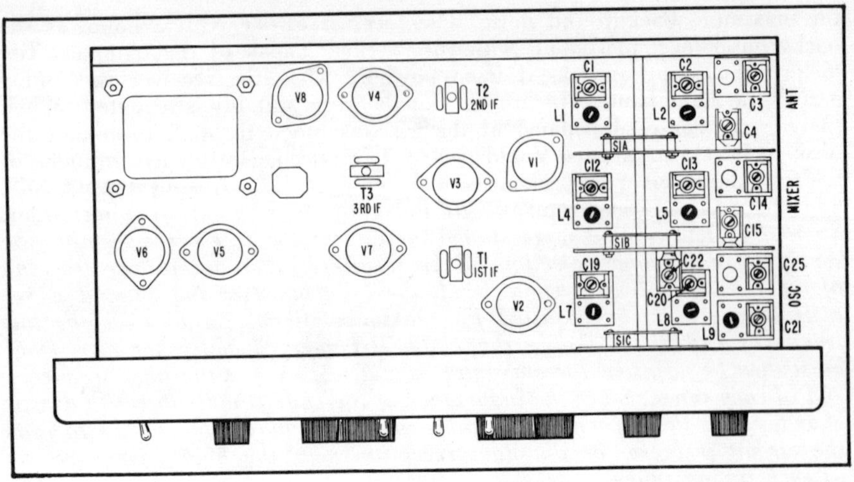

Fig. 8 Under-chassis drawing of Hallicrafters S-85 showing placement of parts.

the enemy receiver. They knew the oscillator was either 455 kc higher or lower in frequency than the channel the message was being sent on. Two such direction finders would give the bearing and range of the receiver for an accurate "fix." A salvo of gun fire would quickly spell the end of the enemy ship. This system worked both ways, and soon all concerned learned how to suppress oscillator radiation in their receivers to prevent detection by the enemy. Your receiver does not require such suppression, and you can hear the high frequency oscillator of your set in a nearby receiver. Tune the auxiliary receiver 455 kc higher or lower in frequency than the dial reading of your set, and you will probably hear the oscillator S-9 plus.

Frequency Calibration of Your Receiver

The dial of your receiver is probably calibrated in terms of frequency. Do not jump to the conclusion that your receiver is as accurate as WWV! It isn't, and it isn't meant to be! Frequency meters designed to check frequencies in the shortwave range cost over a thousand dollars—and are built like a battleship, not like a receiver. Once again, the receiver manufacturer must make a compromise between calibration accuracy and receiver cost. Too many embryo amateurs believe that the dial calibration of their receiver is the Gospel Truth, and adjust their transmitter frequency accordingly. That is why the Federal Communications Commission monitoring stations are working overtime to log the many offenders that trespass outside the legal limits of the amateur bands! The dial calibration of your receiver is a close generalization that will match the majority of a production run of perhaps five hundred or a thousand receivers. If your receiver calibration is "on the nose" across all the bands, you deserve the gold medal with two peanut clusters. When you take the time to examine your receiver calibration closely with the aid of a *good* frequency standard, you will find that the calibration error is not bad—not bad at all, and probably inversely

"As I understand it, the poor chap couldn't decide on which side of the signal to place the local oscillator!"

proportional to the amount of money that you paid for the receiver!

Calibration Errors

If the high frequency oscillator is not detuned from the reception frequency by the proper amount, a simple frequency error will result. If the oscillator is supposed to jog along 455 kc higher in frequency than the received signal, and through some maladjustment it remains only 430 kc away, the dial calibration error will be the error-difference in frequency, or 25 kilocycles. This error will remain constant across the dial, that is, the calibration error will be the same at both ends of the dial.

The smart reader who has digested chapter 4 will remember that the high frequency oscillator tunes at a different *percentage* tuning rate than that of the r-f and mixer circuits. While the oscillator covers the same *number* of kilocycles covered by the r-f circuits, the percentage change from one end of the tuning range to the other as compared to the r-f and mixer circuits is *not* the same. This interesting fact means that the calibration error at one end of the tuning dial can be different from the error at the other end of the dial. It is the frequency of the conversion oscillator that determines the exact frequency of reception, regardless of what markings the tuning dial may happen to have! For example, the dial calibration error might be 25 kilocycles low at the high frequency end of a particular band. At this spot the h-f-o is 25 kilocycles away from where it should be. At the low frequency end of the same band, the dial calibration error might be 375 kilocycles high. The oscillator has strayed 375 kilocycles from the correct frequency. The calibration error can either be high or low at either end of the band, and it can be different at each end! This type of misalignment is called *tracking error*.

The upshot of this whole business is that dial calibration accuracy is almost proportional to the cost of the receiver, and to the care the manufacturer takes in aligning the oscillator circuits. Just because your receiver cost a lot of money do not assume the dial calibrations are correct. They may be no more correct than the calibration of a second-hand "cracker box." *But* the expensive receiver has adjustment points that permit the tracking to be aligned very closely to the dial calibrations with the expenditure of a little of your time. The calibration of the "cracker box" will always be in error, since there is no means of making compensating adjustments. The low cost of the set precluded inclusion of expensive trimmers and adjustment slugs. The luxury of accurate dial calibration costs money, and you must pay for it!

The chief engineer of a large receiver manufacturer once told the author that it cost over twenty dollars of labor time to correctly align one of the company's expensive receivers. This was almost equal to the profit they hoped to make on the receiver! In order to reduce the alignment cost the receivers were adjusted at only *one spot* on each band, near the middle of the dial. The engineer admitted that the alignment job was mediocre and that dial calibration was left pretty much to chance. "It was the only thing we could do," he explained. "Anyway, the receiver was a good one, and the owner who knew the ropes could always adjust the receiver for perfect alignment. We simply could not afford to spend the extra money necessary to provide a first-class alignment job." The receiver in question was a fine one and had plenty of reserve gain. The owner who did not worry about dial calibration probably never noted the "quick-and-dirty" alignment technique used on the receiver. Everyone was happy except the chief engineer. He developed ulcers shortly after the new aligning technique went into use.

High Or Low?

It can be seen that the conversion oscillator of a receiver can be placed either on the low frequency side or the high frequency side of the tuned circuits of the mixer and r-f stage. In the great majority of cases, the high frequency position is chosen. That is, the oscillator is always 455 kilocycles *higher in frequency* than that of the received signal. In a few receivers (such as the *Hammarlund HQ-140* series) the oscillator is placed on the low frequency side of the incoming signal. You can be sure that "high side" operation is employed if series trimmer capacitors are used in the oscillator circuit (C28, C29, C30 of Figure 11, and C23, C24, and C25 of Figure 7). Usually no oscillator series trimmer capacitor is required for the highest frequency band, and the oscillator may be placed on either side of the incoming signal. Many receivers place the oscillator on the "low side" on this band. Unfortunately, this point is seldom mentioned in the instruction manual. Be sure you know on which side of the signal the high frequency oscillator of your receiver operates before you start to play with the alignment trimmers! If you inadvertently "flop sides" and get the oscillator on the wrong side of the signal, you can tweak the trimmers all day and you will never be able to make the oscillator tracking match the dial calibration! Too few instruction books provide this necessary information, which can often only be obtained by writing to the manufacturer of the receiver.

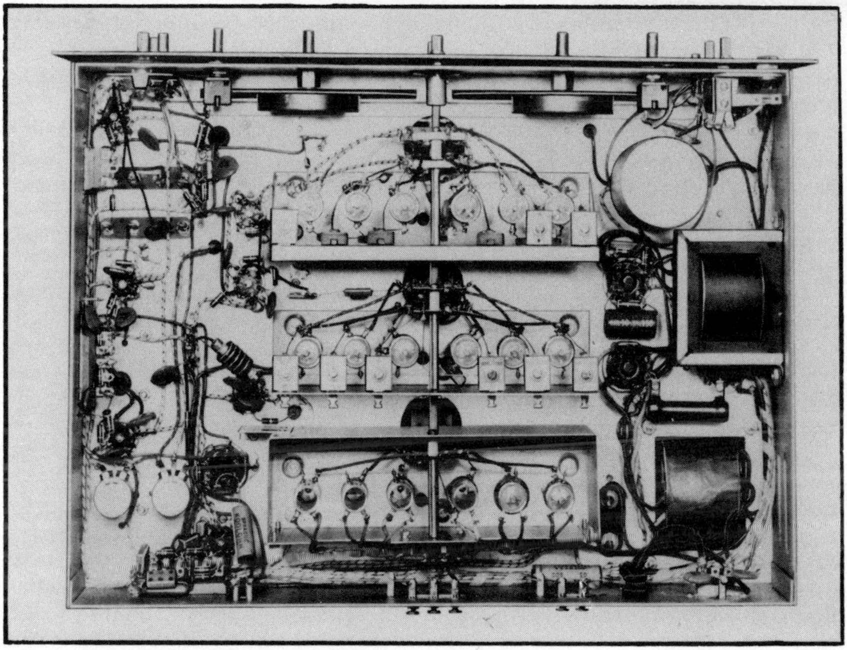

Fig. 9 Under-chassis view of Hammarlund HQ-140XA. High frequency coils are arranged in three banks around bandchange switch at center of receiver. The power supply is at right, with i-f amplifier and tubes along left of the chassis.

Oscillator Adjustment

If you know a few of the landmarks, you can thread your way through the wilderness of receiver alignment with comparative ease. The first landmark is the frequency of the i-f amplifier. This will tell you the displacement of the high frequency oscillator from the signal. Unless warning landmarks indicate to the contrary, you can assume the oscillator is placed on the high frequency side of the signal. Thus if you have a receiver that is far out of alignment, and you know the receiver has an i-f of 455 kc you can easily whip the set into quick alignment. Tune the receiver dial to 15 mc and put an antenna on the receiver. Adjust the oscillator trimmer until you hear WWV. The oscillator is now on 15,455 kc. You can now proceed with a polished alignment job, once you have established yourself with a familiar landmark.

The high frequency oscillator determines the accuracy of dial calibration of your receiver. The tuning range of the oscillator may be adjusted in the same manner as the tuning range of the receiver r-f circuits. That is, oscillator calibration is adjusted first to the low frequency end of each range by varying the inductance of the oscillator coil. The receiver is next tuned

Fig. 10 The frequency control room of NBS station WWV at Beltsville, Md. Fred Sera, Chief Engineer, adjusts r-f generator. Frequency control equipment is located in temperature controlled underground room beneath the station.

to the high frequency end of the dial, and oscillator calibration is adjusted at this point by means of the oscillator trimmer capacitor. The oscillator tuning slugs of the *National NC-188* may be seen in Figure 3. and the high frequency calibration trimmers are visible in Figure 4. Notice that low-drift ceramic capacitors are employed in the oscillator circuit to insure frequency stability with respect to ambient temperature.

The coil catacomb of the *Hammarlund HQ-140XA* receiver is shown in Figure 9. Notice that ceramic capacitors are used on the high frequency oscillator coils for maximum frequency stability, and that all of the mixer and oscillator coils are slug-tuned for tracking adjustment purposes. As with the NC-188, slug tuning of the oscillator coils is required to achieve "low-end" alignment of the r-f circuits.

The circuit of the NC-188 high frequency oscillator is shown in Figure 11. You will notice in the top view of this receiver (Figure 3) that the plates of the oscillator section of the tuning capacitor are not cut to a special shape as are those of the inexpensive ac-dc broadcast set. Such an expedient will work over the broadcast band, but is definitely unsuited for shortwave operation. In order to maintain the desired *percentage* frequency difference between the oscillator and the r-f circuits, a series padding capacitor is added to the oscillator circuit (C28, C29, and C30 of Figure 11). The value of each series capacitor is carefully chosen at the factory to maintain the correct percentage frequency difference as the circuits of the receiver are tuned. Since the percentage rate is the same across a given band, the series padder capacitor may be a fixed unit.

You will notice that this capacitor increases in value for each successively higher frequency band, and the highest frequency band requires no capacitor at all. This is because the percentage frequency difference becomes a smaller factor as the frequency reception is increased, and the reactance of the capacitor may be reduced accordingly. On the highest band, the required capacitive reactance is so small that the capacitor may be omitted with no

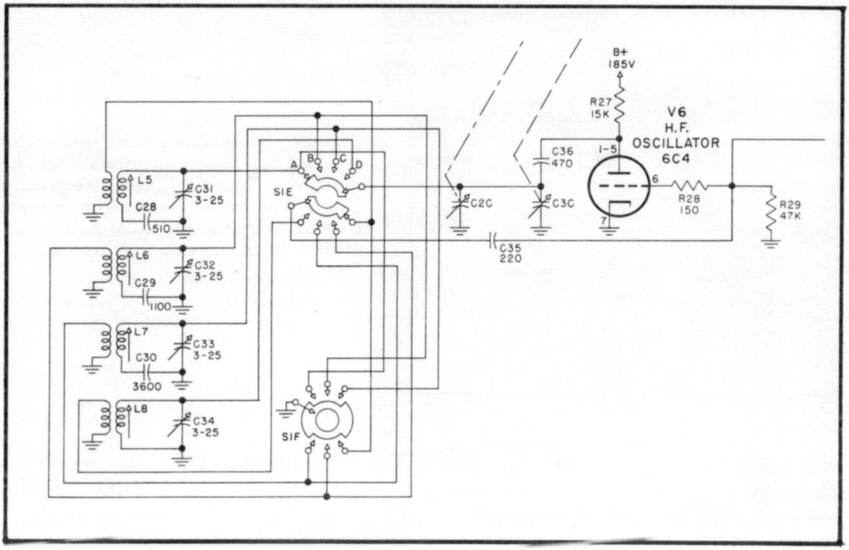

Fig. 11 Oscillator circuit of NC-188 is typical of most receivers. Stability is mainly a function of the mechanical assembly and precision of component parts.

detrimental effect upon receiver performance. Note that the *Hallicrafter S-85* receiver employs a variable series capacitor (C25, Figure 7) for broadcast band tracking. This adjustment is made at 600 kc for correct dial calibration and minimum tracking error.

Complete oscillator alignment should follow the procedure outlined later in this chapter. If the calibration error is not great, slight adjustments may be made to the trimmer and the oscillator coil slug to bring the receiver into calibration. Stations whose frequencies are accurately known may be used as "markers" (WWV, for example), or a crystal calibrator such as the one described in chapter 6 of this Handbook may be used. Low frequency dial calibration is done with the slug of the oscillator coil, and high frequency adjustment is done with the trimmer capacitor. These adjustments tend to be interlocking, and calibration at both ends of the dial should be checked after any adjustment is made. Keep a record of the dial error, and you can soon see if your efforts are making an improvement in calibration.

Complete Receiver Alignment Procedure

If the necessary equipment is at hand, a complete alignment job may be done on your receiver with little trouble. It is an interesting task, and you will learn a lot about receiver operation if you undertake to do this. You will need three pieces of equipment: A signal generator, an output meter, and an alignment tool. A block diagram of the test set-up is shown in Figure 12, and a photo of an actual alignment run on a *Hammarlund*

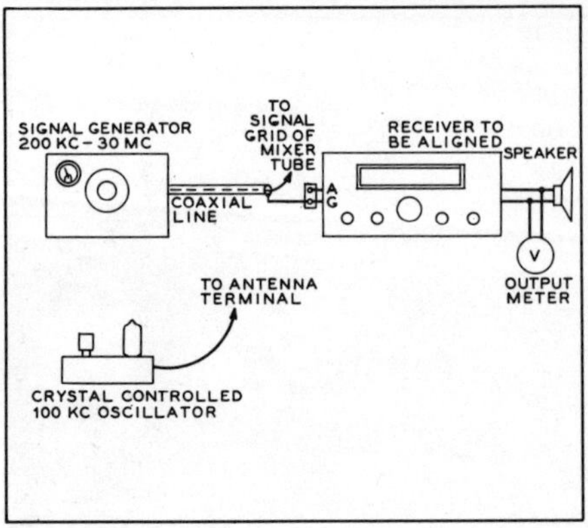

Fig. 12 Tuneable signal generator is used for rough alignment of the receiver. Tracking and frequency check requires accurately known sets of frequencies, most easily obtained from crystal calibrator.

HQ-140XA is illustrated in Figure 13. The vacuum-tube voltmeter used is a *Heathkit V-7A,* and the signal generator is a *Heathkit LG-1*. Other instruments having similar characteristics may be used in their place, if desired.

Preliminary Inspection

Examine all soldered joints. Look for frayed wires, loose screws and nuts. Clean the dust from the components. Apply power to the set, and observe that all tube heaters glow. Look for signs of short circuits and smoke. Measure all the important d-c voltages with the vacuum tube voltmeter and compare these readings with the values given in the receiver instruction book.

Intermediate-Frequency Alignment

Attach the vacuum tube voltmeter (v-t-v-m) to the speaker output terminals, and set it on a low a-c range (0-15 volts). Connect the output of the signal generator to the *signal grid* of the mixer tube through a .01 uf paper capacitor. Ground the shield of the generator coaxial line to the chassis of the receiver. (Note: If the receiver is an ac-dc set, ground the shield of the generator line through a .01 uf paper capacitor to the chassis of the receiver.)

With the signal generator tuned to 455 kc and modulated 30% at 400 cycles, increase the output of the generator to provide an indication on the v-t-v-m output meter. Reduce the range of the meter to 1.5 volts and adjust the output level of the signal generator for half-scale reading of the v-t-v-m. (Note: Check your receiver instruction book for the proper audio signal level at the v-t-v-m. It varies from receiver to receiver.) Tune the trimmers of each i-f transformer (starting with the last) for maximum indication on the output meter. Reduce the r-f output of the signal generator as needed to prevent receiver overload. Recheck alignment of the transformers until you are sure they are carefully peaked. If a crystal filter is incorporated in the receiver, check your instruction manual for alignment information.

Remove the tone modulation from the signal generator, and switch the receiver to c-w operation. Connect headphones to the receiver. Adjust the

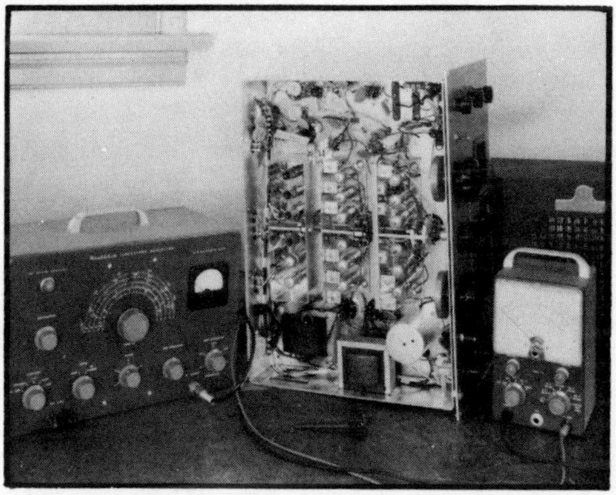

Fig. 13 Typical receiver alignment set-up. Signal generator (Heath LG-1) is at left, connected to antenna terminals of receiver. Vacuum-tube voltmeter (right) is Heath V-7A, attached to speaker terminals of receiver. Tone modulation is used on test signal. I-f checks are done first, then r-f alignment made.

b-f-o trimmer to produce a zero-beat in the phones with the panel pitch control of the b-f-o in center position. Turn pitch control through its range. The beat note should vary smoothly as the control is varied above and below zero beat.

Oscillator Alignment

A crystal controlled marker generator is required for oscillator alignment. The variable frequency *Heathkit LG-1* may be used for r-f and mixer adjustments. The 100 kc crystal calibrator described in chapter 6 may be used as the marker generator for the following tests.

Couple the output of the marker generator to the antenna terminals of the receiver. Check each band of the receiver to see that the marker signals can be heard at 100 kilocycle intervals across the dial. The procedure is to align the oscillator circuits to the correct frequency at the low end of each band by adjusting coil inductance for maximum signal output. The high frequency end of the ranges are then set to frequency by adjusting the trimmer capacitors. These operations are then repeated until both ends of the range are tuned to the correct dial calibration points.

Broadcast band. Tune receiver until the 600 kc marker signal shows an indication on the S-meter. Adjust the oscillator coil slug until the signal peaks at the correct dial frequency. Keep the signal level low to avoid overloading the receiver. Tune receiver until the 1500 kc marker signal is heard. Adjust the trimmer capacitor of the oscillator to peak the signal at the correct dial frequency. Recheck low end calibration and realign coil slug if necessary. Any change in inductance will require a change in trimmer capacitance. Recheck until both ends of range are on the proper settings, always setting the high frequency end of the band last. If a variable oscillator series capacitor is used, it may be adjusted slightly to change the tuning *rate* of the oscillator. This will vary the tuning "distance" between the end points of the band. Always treat the series padding capacitor as a "last resort" adjustment, and try and obtain alignment without touching it.

Shortwave bands. Adjustment of these bands follows the procedure outlined above. Care must be taken to use the correct marker signal, as many of them will be heard as the receiver is tuned across the band. It is a good idea to spot the correct marker signal by zero-beating it with the signal generator. A marker signal of known frequency is chosen near the low frequency end of each band. The *inductance* (slug) of the proper oscillator coil is slowly adjusted in small steps, until the marker signal is heard at the correct point on the dial. It is common practice to choose calibration points about one megacycle in from the ends of the dial to simplify the alignment problem. On the higher frequency bands, trouble may be had differentiating between the marker signal and its image, especially if the oscillator circuits are misaligned. The variable frequency signal generator should be zero-beat with a convenient marker signal and used for preliminary alignment. If the local oscillator of the receiver should be on the *high frequency* side of the signal, the following adjustment must be made: Turn the oscillator slug all the way out. As the slug is turned in, two points will be noticed at which the signal generator may be heard. The *first* signal (slug furthest out) is the correct tuning point. This is true only if the oscillator frequency *decreases* as the slug is inserted within the coil. This tuning sequence may be checked by tuning in the image signal, which should appear at a dial reading approximately 910 kilocycles *below* that of the real signal. If the oscillator operates on the *low frequency* side of the signal, the *second* tuning peak (slug furthest in) should be used.

The high frequency dial calibration should now be set. The signal generator is now zero-beat with a marker of known frequency near the high frequency end of the tuning dial. Standard frequency station WWV can often be used to identify the frequency setting of the signal generator, or to provide a known calibration point for the receiver. The oscillator trimmer capacitor is adjusted to receive the signal at the correct point on the dial. On the high frequency band, it is possible to find two points on the capacitor adjustment at which the signal generator may be heard. One point is the correct one, and the other one is the image point. If the local oscillator of the receiver operates on the "high side" of the signal, use the setting which provides the smallest value of capacitance. This adjustment may be checked by tuning in the image signal, which should appear at a dial reading approximately 910 kilocycles *below* that of the real signal. Conversely, if "low side" operation is desired, use the setting providing the highest value of capacitance. The image signal will now be 910 kilocycles *above* the real signal.

Once the preliminary adjustments have been made, the crystal calibrator marker signals may be used to make final corrections to the dial calibration. Recheck the low frequency calibration point, and reset it if necessary. Any change in the position of the coil slug will require a readjustment of the trimmer capacitor at the high frequency calibration point as these adjustments tend to be interlocking. Recheck until both ends of the range are on the proper dial settings, always setting the high frequency calibration point last. It should be possible to maintain a calibration accuracy of greater than 0.5% at these points. That is, the calibration error should be less than 150 kilocycles on the main tuning dial at 30,000 kc. With care you can reduce it below this figure.

Be sure and let the receiver run for an hour or two before you align the oscillator. Check the frequency drift caused by thermal heating of the

"Here's yer shortwave set, Buddy!"

chassis and components. It may be necessary to retouch the adjustments to bring the calibration back into line. In any case, it is smart to do the oscillator alignment job after the receiver has reached operating temperature.

R-F and Mixer Stage Alignment

The next job is to adjust the mixer and r-f tuned circuits so that they track properly with the oscillator. These circuits are tuned to the frequency indicated on the receiver dial, whereas the oscillator circuit has been adjusted 455 kilocycles higher or lower in frequency than the dial calibration. The variable signal generator may be used as a signal source, since frequency accuracy is not required.

Alignment procedure is the same for all bands. The signal generator is tuned to the low frequency end of the band, and is adjusted to provide a half-scale reading on the S-meter of the receiver. The inductance of the r-f coil and the mixer coil is varied by means of the coil slug for maximum signal indication on the meter. The signal generator is then tuned to a spot near the high frequency end of the band, and the r-f and mixer trimmer capacitors are adjusted for maximum signal strength. Retune the receiver to the low frequency end of the dial, and note the position of the r-f and mixer trimmers. Repeak the trimmers for maximum received signal at this frequency and note the new setting for both. If the mixer and r-f trimmers do not peak at the same setting in the above two tests, vary the inductance of the coils slightly and repeat the test. It should be possible to have both r-f and mixer trimmers peak at approximately the same setting for both ends of the band. Be sure that you do this test on the actual signal, and not upon the image signal. It can be noticed that the r-f and mixer trimmers will produce two peaks in the background noise on the higher frequency bands. One peak is the correct one, the other is the image peak. On the latter, the tuned circuits are peaked on the incorrect side of the high frequency oscillator. If the oscillator operates on the high frequency side of the incoming signal, the correct noise peak is the one that occurs at the greater capacity setting of the trimmers. Conversely, if the oscillator operates on the low frequency side of the incoming signal, the trimmers should be aligned on the noise peak that occurs at the lesser capacity. When alignment is completed, the image signal will appear 910 kc *below* the real signal if the oscillator is operating on the high frequency side of the incoming signal, and 910 kc *above* the real signal if the oscillator is operating on the low frequency side of the incoming signal.

Errors in Tracking

After the high frequency oscillator, r-f amplifier, and mixer circuits have been aligned, tracking at any point in the spectrum may be checked by adjusting the signal generator to the proper frequency and testing the settings of the r-f and mixer trimmers for maximum gain. Frequency calibration may also be checked at these points. After such a test, all trimmers should be reset at the high frequency end of the band since their settings are most critical at this point.

Tracking errors at the low frequency end of the band can be caused by errors in the tuning capacitor sections, the value of the coil inductance, or the high frequency oscillator series capacitance. In order to determine if one or more sections of the master tuning capacitor are the cause of mistracking, it is necessary to make the check described above on two or more different bands. If the same tracking error appears on all bands, the master tuning capacitor is at fault. The error may be corrected by permanently bending the slotted rotor plates of the capacitor to provide the proper amount of correction.

If the tracking error appears only in the r-f amplifier or mixer stage, and only on one band, the inductance of the coil of the particular stage is at fault or the series padding capacitor is an incorrect value. Some oscillator coils have a moveable turn or loop that may be varied slightly to correct the inductance. Others may be corrected by movement of the variable slug mounted within the coil.

The series padding capacitor for the oscillator (in the great majority of receivers) is a hand-picked low-drift mica or ceramic unit that is not adjustable. If it seems certain that this capacitor is in error, it should be carefully removed from the circuit and measured on a capacity bridge. Replacement units may be obtained from the receiver manufacturer or his accredited service organization.

How Good Is My Receiver?

This argument rages at all radio club meetings, among the amateurs on the air, and at the counters of the radio stores. In fact, it is one of the general topics of conversation whenever two radio enthusiasts get together. Each receiver manufacturer has his own enthusiastic group of exhuberant

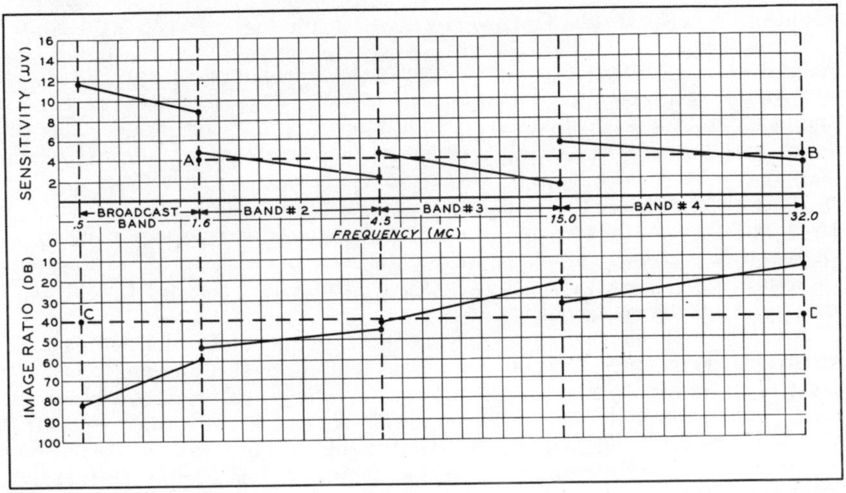

Fig. 14 Sensitivity and Signal-to-Image Ratio is shown above for the typical all-wave receiver of the $150 price range. Sensitivity is adequate over the entire range, but S/I ratio is poor above 5 mc. Image signals are troublesome, particularly on the 20, 15, and 10 meter amateur bands. An extra r-f stage (preselector) will do much to reduce the image signals and improve the signal to noise ratio and sensitivity above 14 mc. The image problem also may be substantially eliminated by the double conversion receiver. The low sensitivity is an asset on broadcast band, as it helps to reduce overload from strong stations.

boosters, and a small tightly-knit clique of knockers. The cherished receiver of one ham is the pet gripe of his best friend. "Never buy a *Floppola-8!* They are a pile of junk. You can't hear *anything* with one of those lemons," says one ham. *The Floppola-8?* Best receiver I ever had," says another. "Used it for my weekly sked with Inner Slobbovia for two years and never missed a word!"

The truth usually lies somewhere between these two violent extremes. When you remember that the original signal plucked out of the air is of the order of a few microvolts, *receiving anything at all* seems like a large order. In general, the old axiom, "you get what you pay for" certainly applies to receivers. Your pocketbook is generally a good guide to receiver performance.

Radio manufacturers are somewhat loath to release specific information about their products, and rightly so. Detailed information would not mean much to the average buyer, and it is a function of the type and technique of measurement. Thus information produced by one manufacturer would not necessarily be based upon the same measurements and operating conditions used by another. Comparison of the two groups of information, therefore, would be meaningless. Then too, many manufacturers shy clear of talking too much about the specific performance of their products in order to protect themselves against other less scrupulous manufacturers who dream up their performance characteristics in the advertising department, rather than in the engineering office.

Typical Receiver Performance

With little trouble it is easy to reach a broad plateau of agreement concerning receiver specifications. These "specs" are deliberately vague, but they can provide a yardstick by which a receiver may be judged. The two basic points of interest to all receiver owners are *receiver sensitivity* and *signal to image ratio* (S/I). These points do not define all the excellent characteristics or all the faults of a receiver. A certain set can excel in these features, yet have drift, hum, and poor overload performance in the presence of a strong signal. However, this is a good place to start, so let's look at a typical receiver in the $150 price range and see just how it measures up to these requirements.

An arbitrary performance specification might state that satisfactory receiver performance can be defined as entailing a minimum sensitivity level of 4 microvolts and a signal-to-image ratio of 40 decibels. The sensitivity specification refers to the minimum value of input signal required to produce an audio signal that is 10 decibels above the noise of the receiver. Taking these specifications at their face value, we can apply them to the mythical receiver under examination. This receiver is a nine tube set, having one r-f stage and a 455 kc i-f amplifier. It covers 550 kc to 32.0 mc in four tuning ranges. The sensitivity and S/I ratio curves are shown in Figure 14. Sensitivity is deliberately reduced on the broadcast band to about 10 microvolts to eliminate cross-talk and overloading that might be caused by strong local stations. Extreme sensitivity is not required for this band as local atmospheric noise will mask the receiver sensitivity in any case. Image rejection is best at the low frequency end of the band, being over 80 db. The image ratio drops to about 60 db at the high frequency end of the broadcast band. This deterioration in S/I ratio is normal, and is a function of the ratio of inductance to capacitance in the tuned circuits at any given frequency. You can see from Figure 14 that the best S/I ratio occurs when the receiver is tuned to the low frequency end of each band.

The sensitivity in each shortwave band is acceptable, being better than 6 microvolts on each band. This is sufficient to allow the listener to hear the "background noise" picked up by the antenna at all frequencies except perhaps those above 25 mc or so. The external noise level on the 10 meter (28—29.7 mc) amateur band is quite low and signals having strengths of the order of 0.5—5.0 microvolts can be heard. This particular set will probably miss these "real weak ones." Notice that the sensitivity of the receiver is at maximum at the high frequency end of each tuning range. This is because the most favorable coupling conditions between the r-f coils of the receiver occur at minimum capacity setting, and is one reason why the manufacturers like to place each amateur band so that it falls at the top end of each tuning range. This insures that the receiver will be "hot" in the ham bands.

If we hold to our concept of a signal-to-image ratio of 40 db as being satisfactory performance, we see that this particular set falls below that requirement above 5 megacycles. This means that image signals of strong 40 meter amateurs can be found on the receiver by looking for them, that images of 20 meter signals will be apparent, and that images of 15 meter amateurs will be very noticeable. Images of 10 meter amateurs will be almost as loud as the fundamental signal, being reduced in strength by only about 15 db (3 S-units).

This is an unpleasant fact of life, but one that must be faced. A receiver in this price class usually has two tuned circuits (r-f and mixer) preceding the i-f stages. It is up to these tuned circuits to separate the real signal and the image signal, which are separated by 910 kilocycles. Two tuned circuits can do a pretty good job of separation at a frequency of, say 4,500 kilocycles. The *ratio of separation* to that of the resonant frequency is 910/4500, or 20.2%. When the circuits are tuned to 30,000 kilocycles the ratio of separation drops to 910/30,000, or only 3.03%. Two tuned circuits, such as the ones used in receivers of this type just cannot offer any degree of discrimination to two frequencies that differ by such a small percentage. As a result, the image problem at 10 meters is bothersome. It can be seen that the degree of image rejection is proportional to the ratio of the received signal to the image frequency, which is also proportional to the ratio of the received signal to the intermediate frequency of the receiver. Generally speaking, when commercial grade components are used, image interference becomes a problem above 10 mc in a receiver having a 455 kc intermediate frequency amplifier and only one r-f stage.

What to Do?

There are several solutions to the image problem. The first and least expensive is to add an additional stage of r-f amplification to the receiver. An extra tuned circuit or two will do wonders to improve the S/I ratio. Some of the more expensive receivers have two r-f stages to help overcome the image problem. The addition of an extra stage, however, is a "brute force" solution to the problem. A more subtile approach is to raise the intermediate frequency to increase the ratio of separation figure. An intermediate frequency of 3,000 kilocycles will provide a ratio of 20% at 30 mc and will increase the S/I ratio enormously. An i-f amplifier operating at this frequency has very poor selectivity and low gain, so it is necessary to use a second mixer tube to convert from this intermediate frequency down to a lower frequency to pick up gain and selectivity. A receiver that accomplishes these functions is called a *double conversion* receiver, and its outstanding attribute is relative freedom from bothersome image signals.

Another solution is to employ a converter with your receiver. The combination of receiver-plus-converter is equivalent to a double conversion receiver in that it combines good image rejection and high selectivity. Converters are often used with high quality communications receivers to extend their range above 30 mc. (A case in point: the *National NC-300)*.

An excellent check for receiver sensitivity is to remove the antenna (repeaking the antenna trimmer if necessary) and compare the set noise with and without the antenna. If the receiver can "hear" the atmospheric and antenna noise, the internal noise level of the receiver is low enough for weak-signal reception. In any event, the use of a *good* receiving antenna will improve the performance of even a relatively "flat" receiver.

For the amateur or SWL who already owns a receiver that is subject to image interference, the simple preselector described in the next chapter will do much to overcome this problem, and at the same time will provide increased sensitivity over the high frequency portion of the spectrum.

CHAPTER VI

Receiving Equipment You Can Build

Part of the fun of any scientific hobby is the thrill of building your own equipment. The "know-how" obtained by constructing a piece of electronic gear cannot be duplicated by reading a thousand books! The personal satisfaction gained from the assembly of your own "brain child" is worth much more than the small cost of the parts needed to make the unit. Shown in this chapter are several interesting projects that are easy to build, and which will make worthwhile additions to your listening post. Thrown in for good measure are some hints and ideas concerning the fine art of equipment construction.

Armed with a few basic tools, you can build almost any piece of radio equipment that you desire. The enjoyment of using the equipment is second only to the fun that you will get out of completing a "do-it-yourself" project. You can do all the work, including the layout and drilling of the chassis, or you can perhaps buy the item you desire in kit form wherein all the metal work has been done for you. It is possible, however, to have almost any metal work done for you at nominal cost in a small sheet metal shop. As you progress, you will probably collect the necessary tools to do all but the most elaborate sheet metal work yourself.

Whether you build from a packaged kit, or "roll your own," you must be adept in the use of the soldering iron, and know how to make good, rosin-free joints. A few words on this subject would not be amiss.

Proper Soldering Procedure

Correct soldering technique is extremely important. Good soldered joints are essential if the performance engineered into a piece of radio equipment is to be fully realized. If you are a beginner with no experience in soldering, a half-hour's practice with odd lengths of wire, a tube socket, and some components will be a worthwhile investment.

High quality rosin solder of the proper grade is most important. There

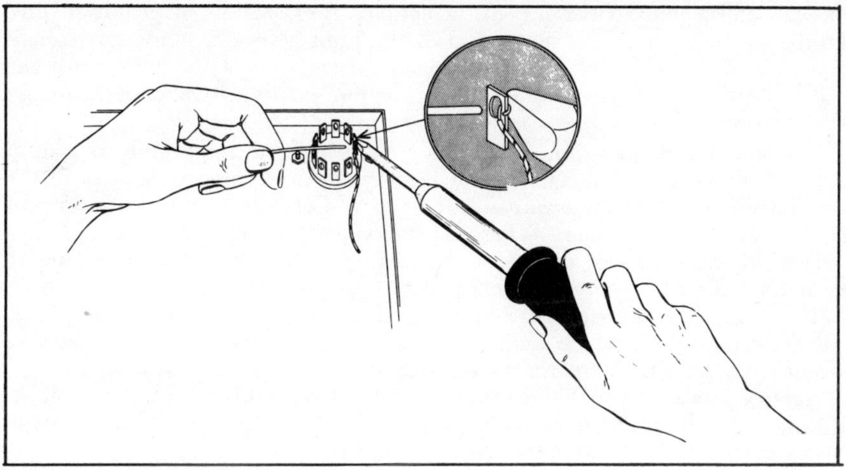

Fig. 1 Proper soldering technique is necessary for construction of equipment. Soldering iron is used to heat the work sufficiently to melt the solder (above).

are several different brands of solder on the market, each marked "Rosin Core Solder." Such solders consist of an alloy of tin and lead, usually in the proportion of 50:50. Minor variations exist in the mixture, such as 40:60, or 45:55, etc., with the first figure indicating the tin content. Radio solders are formed with one or more holes running lengthwise through the center. These holes are filled with a rosin compound which acts as a flux or cleaning agent during the soldering operation.

No separate flux, acid, or cleaning agent of any kind should be used in radio work. Such compounds, although not always corrosive at room temperatures, will form residues when heated. The residue is deposited on surrounding surfaces and attracts moisture. The resulting compound is not only corrosive but actually destroys the insulation value of non-conductors. Dust and dirt will tend to accumulate on these surfaces and eventually will create erratic or degraded performance of the equipment.

If terminals are bright and clean and wires free of wax, frayed insulation and other foreign substances, no difficulty will be experienced in soldering. Crimp or otherwise secure the lead to the terminal so that a good connection is made without relaying on solder for physical strength. To make a good-solder joint, the clean tip of the soldering iron should be placed against the joint to be soldered so that the terminal is heated sufficiently to melt solder (Figure 1). The solder is then placed against both the terminal and the tip of the iron and will immediately flow out over the joint. Use only enough solder to cover wires at the junction as it is not necessary to fill the entire hole in the terminal with solder. Excess solder may flow into tube socket contacts, ruining the socket, or it may creep into switch contacts and destroy their spring action. Position the work so that gravity tends to keep the solder where you want it.

A poor solder joint will usually be indicated by its appearance. The solder will stand up in a blob on top of the connection, with no evidence of flowing out caused by actual "wetting" of the contact. A crystalline or grainy texture on the surface, caused by movement of the joint before it solidified is good evidence of a "cold" connection. In either event, reheat the joint until the solder flows smoothly over the entire junction, cooling to a smooth, bright appearance.

A good, clean well-tinned soldering iron is also important to obtain consistently perfect connections. For most radio work, a 60 watt iron, or the equivalent in a soldering gun is satisfactory. For tight corners, a "Pencil-type" iron is a great help. When many connections must be soldered to a copper or steel chassis, a 300 or 400 watt iron should be used to insure even heat distribution over the chassis.

For additional information concerning radio components, tools, and helpful construction hints, the reader is referred to the *Novice and Technician Handbook*, published by Radio Publications, Inc., Wilton, Conn. This informative book is available through your local radio distributor. It contains a gold mine of useful information for the home constructor in addition to many exciting electronic construction projects.

Radio Tools

Many hams can build acceptable radio equipment, armed only with an icepick and a dull Boy Scout knife. This practice is not to be recommended. Rather, it is a wise idea to provide yourself with a good set of tools that will enable you to do the maximum amount of work with the minimum amount of effort. Here's a suggested list of basic tools used most often in building electronic equipment. Buying them one at a time will not deflate the pocket book too fast, and in a short time you will have a first class radio laboratory.

1. A *good* 100-watt soldering iron or soldering gun. The *Weller model 8100K* gun, or the *Drake #325 iron* or equivalent are recommended. The iron will cost $3.75, and the gun about $6.00.
2. Solder. A one pound spool of *rosin core* solder is recommended. Don't buy less than a pound, as solder is very expensive in small amounts. A pound roll will cost slightly over a dollar. The smart reader will read the soldering notes later in this chapter before he rushes out to buy this all-important item.
3. Six-inch long nose pliers, such as the *Kraeuter 1671*. Cheap pliers are worse than useless, so watch out for "bargains". Good long nose pliers will cost about $2.00.
4. Six-inch side cutters, such as the *Kraeuter 1830*. These cost slightly more than the pliers, and are worth their weight in gold. Don't try to do without them!
5. Screwdrivers. You need two, one with a six-inch shaft, and one with a three-inch shaft. You should make sure these tools have insulated handles, and that the tip of the blade is square and true. A suggested tool is the plastic handled *Xcelite* screwdriver, available in most radio stores. Steer away from the heavy metal-handled "automobile type" screwdriver. You're not building a tractor!
6. A good pair of pliers. Snoop around your automobile accessory store for well-made pliers. They should have very little side play in the joint. The *Kraeuter Grip-tite 356* or equivalent fills the bill nicely.
7. An assortment of files . . . flat, half-round and rat-tail will come in very handy. Your local five-and-dime store will probably have a fine collection for you to pour over.
8. A pistol-grip hack saw and assorted blades. Very important for chassis work!
9. A good carpenter's brace (or a sturdy hand drill) and an assortment of drills. One each 1/16", 1/8", and 1/4" will do for a starter.

This, then, is just about the absolute minimum of tools you will need to set up your shop. The next most important items to procure are a hammer, center punch, nut-driver, assorted chassis-punches, an electric hand drill, and a wire-stripper.

Tools and More Tools

The soldering iron is one of the basic tools of the trade. Contrary to popular opinion, it is not used to melt solder, rather it is used to heat the part to be soldered to the point where it will melt the solder. While the soldering iron is heavier than a gun, it will supply heat whenever you want it, after the initial warmup period. The soldering gun is lighter and consumes less power. For general use, the iron is to be preferred, although a soldering gun is a mighty handy tool to have around. Buy 'em both if your pocketbook can stand it.

Side cutters are used to cut wire, and to strip insulation off the end of the wire. Practice removing insulation off the wire with these. Soon you should be able to remove about one-half inch of insulation from the end of the wire without nicking the conductor, or cutting it in two.

After you have advanced in the fine art of electronics, you may wish to purchase other tools. By now you have found that some equipment uses foxy cross-slotted screws and bolts. A *Phillips head* screwdriver will take care of most of these in short order. You will also find that a good electric drill (in lieu of the hand drill) will save a lot of wear and tear on your frame! A three-inch vice is also a handy item to have at hand.

Test Equipment

You can't get very far without a few basic measuring devices for electrical currents and voltages. As in the case of hand tools, "you get what you pay for". If you are adept at soldering and construction, you can save a tidy sum by buying your electronic instruments in kit form, and assembling them yourself. Otherwise, you must pay more for the equivalent in finished form.

The first instrument to buy is the volt-ohmmeter which forms the backbone of your test equipment. This handy device will measure a-c and d-c voltages, direct current and resistance. A pocket-size volt-ohmmeter will cost from twenty to twenty-five dollars. In kit form, it will cost about fifteen dollars or so. A good instrument should have an insulated case, and the meter should have a zero-set adjustment. Beware of "off-brands" and "special bargains". If you stick to name brands you will not go wrong. False economy when buying test equipment is costly to the pocket book. Buy the best you can afford.

The next item of test equipment to purchase is an r-f signal generator covering the range of 200 kc to 30 mc. Several such generators are available in kit form, and may be purchased for under thirty dollars. A signal generator is useful in aligning radio receivers, and is a "must" item in a well equipped shop.

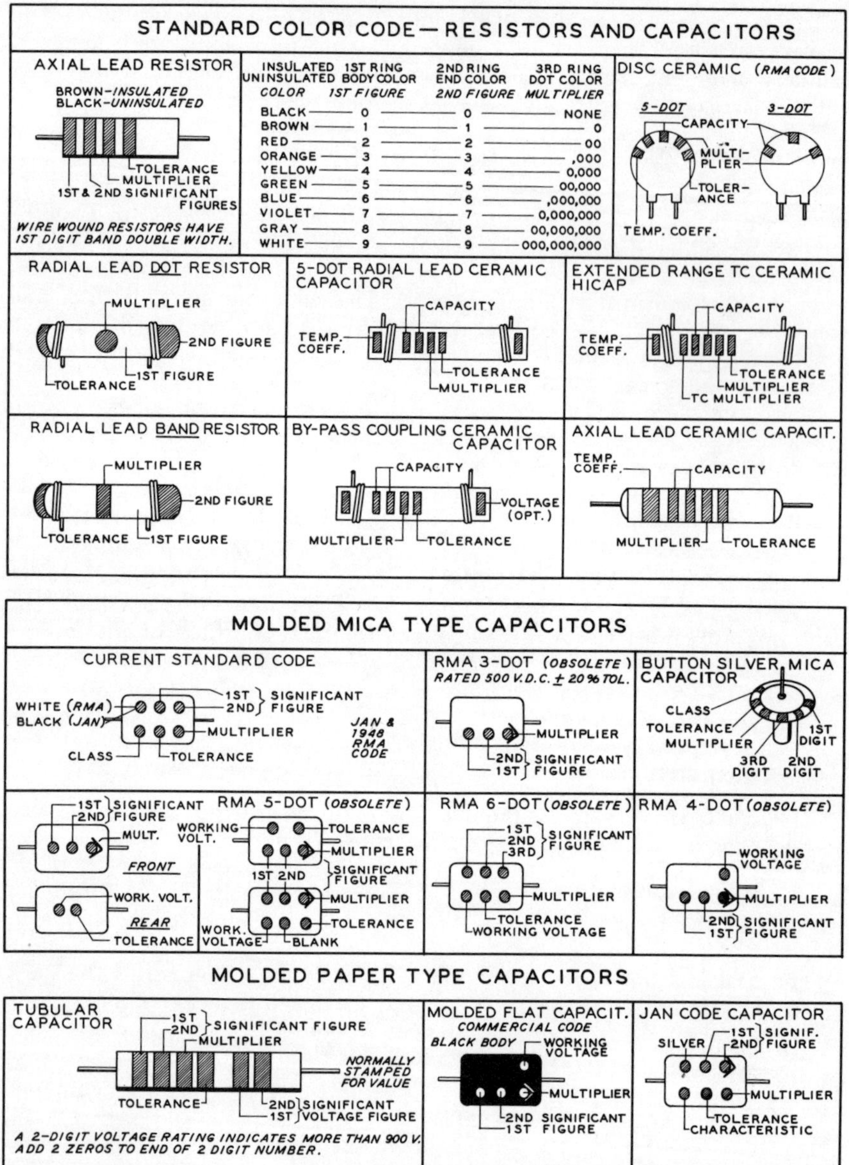

Fig. 2 Color code is used to indicate the values of resistors and capacitors. Code is read from left to right, with first color band indicating the first digit, the second color indicating the second digit, and the third color indicating the "multiplier." For example, a resistor coded yellow-violet-green has "four" for the first digit, "seven" for the second, and a multiplier of "five," or "00,000." The value therefore is 4,700,000 ohms, or 4.7 "megohms." Memorize this code!

Sources Of Supply

Every large city has at least one wholesale radio distributor where the enthusiast may buy parts and equipment at prices considerably below the so-called catalog "list price." Many of these stores have a mail order department to handle out-of-town orders. It is a smart idea to go through the advertising section of an electronics magazine and note a few of the radio distributors advertising therein. A postal card will bring you their free catalog which can provide you with price information and a wealth of data about components and finished equipment. Careful reading of these catalogs enables you to identify parts and components that are unfamiliar to you.

All the components required in the equipment described in this chapter may be purchased from these catalogs or may be bought directly from a radio distributor. A complete parts list for this purpose is provided with each unit.

A Crystal Calibrator For Your Receiver

What frequency is your receiver tuned to? Your dial calibration is only an approximation! Knowing the exact frequency of reception is an invaluable help in DX-hunting! The amateur operator must know the frequency limits of the ham bands to prevent off-frequency operation of his transmitter. The addition of a one-tube crystal calibrator to the receiver will provide you at the flip of a switch with accurate marker signals every 100 kilocycles across the dial. Build this inexpensive, precision unit as first step towards better listening. Cost of parts about $10.

This simple frequency calibrator is a precision, crystal controlled oscillator that may be used with any communication receiver *not* of the ac-dc type. The frequency of oscillation is held very closely to 100.00 kilocycles, and can be placed "on the nose" by means of the calibration control capacitor C1, seen in the schematic diagram. It is interesting to note that your receiver doesn't tune to 100 kc, and you will never hear the signal of the calibration oscillator! But you *will* hear the harmonics (overtones) of the oscillator every 100 kc up the radio spectrum, all the way to 30 mc or so. The frequency of the harmonics is regulated by the fundamental frequency of the crystal oscillator, and when the unit is properly adjusted, these "lighthouses" will serve you well at every 100 kilocycle interval. Your receiver cannot tune to the lower harmonics (200, 300, and 400 kc), but you can hear the fifth harmonic at 500 kc, the sixth at 600 kc, and so on. The harmonic signals continue at these intervals across the complete tuning range of your receiver, acting as reliable marker points at known intervals throughout the entire span of the dial.

It might seem hard to identify a particular calibrator harmonic in the shortwave spectrum. For example, how do you tell the 16,500 kc marker signal from the 16,600 kc marker, or even the 16,400 kc marker? The one you desire does not have a green strip painted down its back! The solution

Fig. 3 A crystal calibrator oscillator may be used to provide the marker signals every 100 kilocycles across the shortwave spectrum. Shown above is such an oscillator mounted within a Hammarlund HQ-140XA. Voltages to run the unit may be "borrowed" from the receiver with no ill effects. The oscillator is controlled by the toggle switch mounted on top of chassis. Crystal is at right.

to this problem of identity is to employ the transmissions of Standard Frequency Stations WWV and WWVH in conjunction with the calibrator. The 2.5, 5.0, 10.0, 15.0, and 20.0 mc harmonics of the crystal calibrator will coincide with the marker signals of these stations. When the standard frequency is found, your desired harmonic may be spotted by noting the number of calibrator harmonics between marker signal and your harmonic.

The Crystal Calibrator Circuit

The schematic of the calibrator is shown in Figure 4. A miniature 6BA6 pentode tube serves as a crystal oscillator. The frequency of oscillation may be varied a small degree by adjusting calibration capacitor C1. Switch S1 controls the plate voltage to the unit, permitting the calibrator to be turned off without removing the filament voltage from the tube. The plate and filament voltage requirements are very modest and the necessary power may be "stolen" from your receiver with no strain or pain. The calibration signal is coupled to the antenna circuit of your receiver from the plate circuit of the 6BA6 tube through coupling capacitor C5.

Calibrator Construction

The drilling layout of the crystal calibrator chassis is shown in Figure 5. The entire unit is built upon one-half of a two-piece aluminum box measuring 1½" x 2⅛" x 3½" in size. Placement of the major components may be seen in the photographs. The following wiring and assembly instructions are presented in a step-by-step sequence to enable you to complete the project easily and correctly. Be sure to read each step all the way through before you start to do it. When the step is completed, check it off in the space provided. Do not solder any connections until told to do so. Ready? Let's get going!

( ) Lay out the chassis holes (Figure 5) to be drilled on the paper wrapper of the box. Drill the holes and remove the paper. Open the box and clean off the drill burrs with a file.

( ) Mount the miniature 7-prong socket V1, using 4-40 hardware. Orient the socket so that pins 1 and 7 face the crystal socket holes.

( ) Mount the crystal socket X1, using 4-40 hardware. Place a grounding lug (soldering lug) beneath the retaining nut adjacent to pin B. Be sure the lug does not touch pin B.

( ) Mount the ceramic calibration capacitor C1, using ⅝-inch long 4-40 bolts. Space the capacitor away from the chassis with ¼-inch metal spacers slipped over the bolts. The screwdriver adjustment of C1 should be accessible through the chassis hole.

( ) Mount the off-on switch S1, using the nut and washer supplied with the switch. The contact arms of S1 are oriented towards the crystal socket.

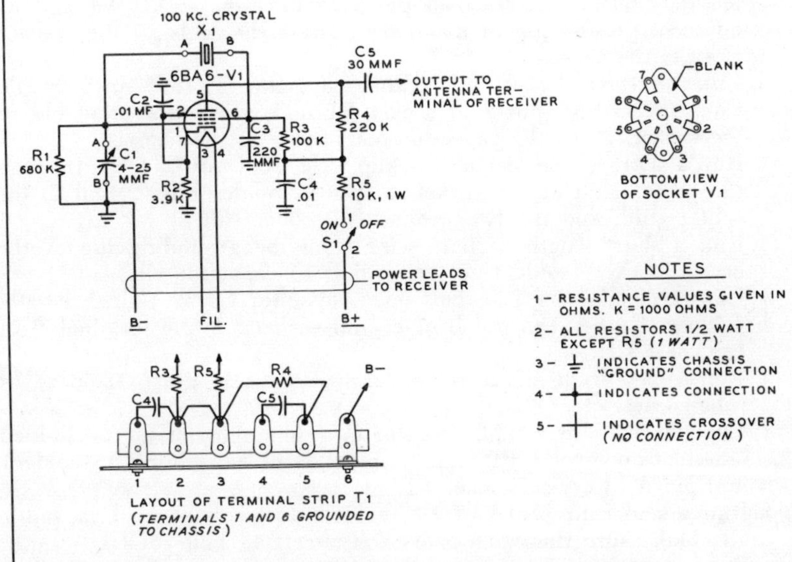

FIGURE 4 – SCHEMATIC, CRYSTAL CALIBRATOR

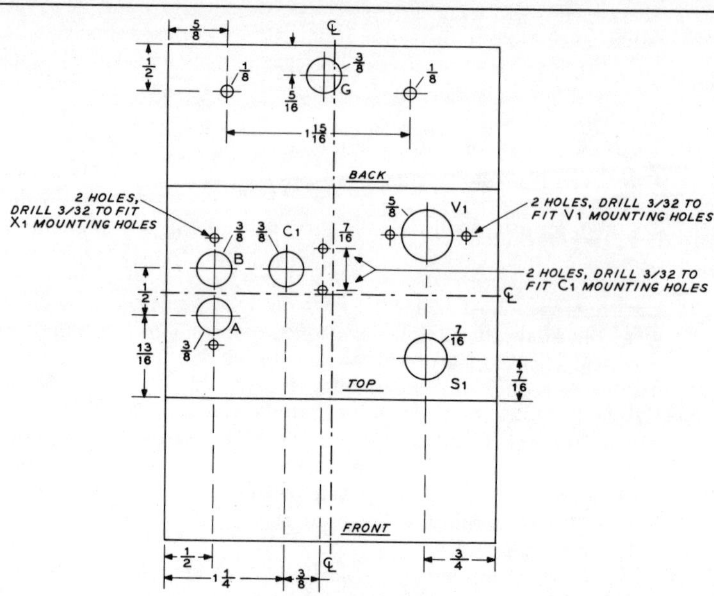

FIGURE 5 — DRILLING TEMPLATE FOR CRYSTAL CALIBRATOR CHASSIS

( ) Place a 3/8-inch rubber grommet G in the side hole of the chassis.

( ) Clip the leads of a 3.9K resistor R2 (orange-white-red) to 3/8-inch length. Connect R2 between pin 7 of the tube socket V1 and an adjacent ground lug of the socket. Dress the body of the resistor close to the chassis.

( ) Clip the leads of a .01 mf ceramic capacitor C2 to 3/8-inch length. Connect between pin 2 of socket V1 and adjacent ground lug of the socket. Dress the capacitor body close to the chassis.

( ) Run a short length of bare hookup wire from pin 2 of V1, through the center stud of the socket, to pin 7. Solder pins 2 and 7, the center stud, and the two ground lug connections.

( ) Run a short length of bare wire from the grounding lug of the crystal socket (solder) to terminal B of capacitor C1.

( ) Clip the leads of a 220 mmf mica capacitor C3 to 3/8-inch length. Connect C3 between pin B of crystal socket X1 and terminal B of capacitor C1.

( ) Run a lead from pin B of crystal socket X1 to pin 6 (solder) of tube socket V1.

( ) Clip the leads of a 620K resistor R1 (blue-red-yellow) to 1/2-inch length. Connect R1 between terminal B of capacitor C1 (solder) and pin A of crystal socket X1.

( ) Run a lead from pin A of X1 (solder) to terminal A of capacitor C1. Make sure this wire clears the mounting bolts of C1.

( ) Run a wire between terminal A of C1 (solder) and pin 1 of socket V1 (solder).

( ) Install the six terminal phenolic strip T1, using 6-32 hardware. Make sure that it does not touch other components.

( ) Clip the leads of the .01 mf ceramic capacitor C4 to 3/8-inch length. Connect C4 between tie-point terminal 1 (solder) and terminal 2 of strip T1.

( ) Clip the leads of the 30 mmf coupling capacitor C5 to 1/2-inch length. Connect C5 between terminals 4 and 5 of strip T1.

( ) Clip the leads of the 100K resistor R3 (brown-black-yellow) to 1/2-inch length. Connect R3 between pin B of crystal socket X1 (solder) and terminal 2 of strip T1.

( ) Run a short length of bare hookup wire between terminal 2 (solder) and terminal 3 of strip T1.

( ) Clip the leads of a 10K, 1-watt resistor R5 (brown-black-orange) to 3/4-inch length. Connect R5 between terminal 3 of strip T1 and terminal 1 (solder) of switch S1.

( ) Clip the leads of a 220K resistor R4 (red-red-yellow) to 1/2-inch length. Connect R4 between terminal 3 (solder) and terminal 5 of strip T1.

( ) Run a short length of bare wire from terminal 5 (solder) of T1 to pin 5 (solder) of tube socket V1.

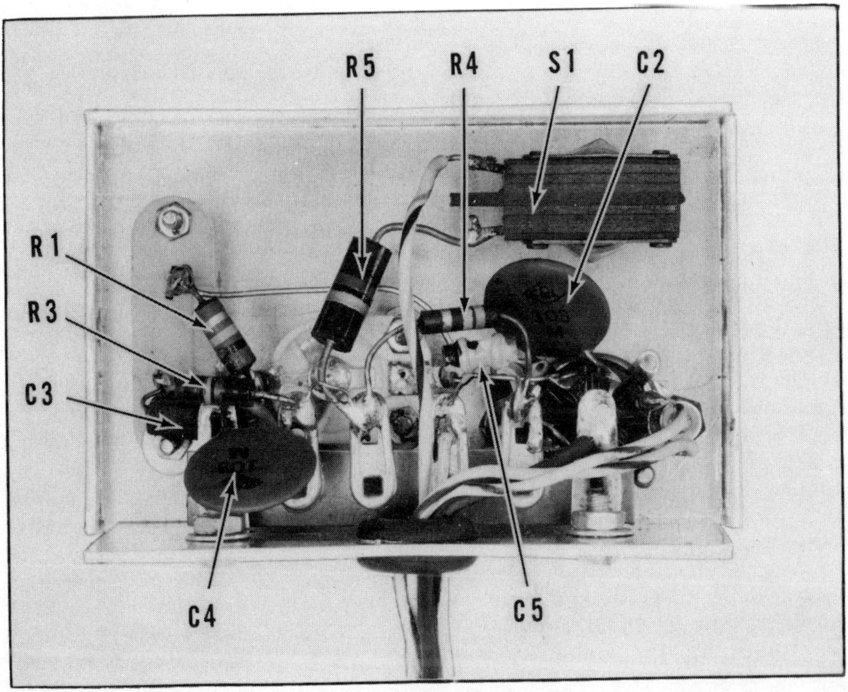

Fig. 6 Under-chassis view of crystal calibrator showing placement of major components. Tube socket is at lower right, and crystal holder is at the left. All small parts are mounted between phenolic terminal strip and major components.

( ) Cut four lengths of hookup wire, each two feet long. Strip ¼-inch of insulation from the ends of the wires, and tin the ends. Wrap a bit of adhesive tape around the insulation at one end of each lead so you can write identifying symbols on the leads.

( ) Take one lead and solder the untagged end to terminal 6 of strip T1. Mark the opposite end of the lead *B-minus.*

( ) Take a second lead and solder the untagged end to terminal 2 of switch S1. Mark the opposite end of the lead *B-plus.*

( ) Take the two remaining leads and twist them together for their full length. Solder the untagged ends of the twisted pair to pins 3 and 4 respectively of tube socket V1. Mark each lead *filament.*

( ) Gather the leads into a cable and pass them through the rubber grommet G mounted on the back of the chassis. Take short bits of adhesive tape and tape the leads into a cable at four-inch intervals.

( ) Cut another two-foot length of hookup wire. Strip and tin the ends. Attach one end to terminal 4 (solder) of strip T1. Mark this lead *output.* Pass the lead through grommet G.

( ) Examine all connections to make sure they are well soldered. Make sure that the unit is free of wire clippings and drill burrs that might create accidental short circuits. Check your wiring against Figure 4.

( ) Attach the bottom half of the aluminum box with ¼-inch 6-32 screws. Place the 100 kc calibration crystal and the 6BA6 tube in their respective sockets. The crystal may be inserted either of two ways, since it has no polarity.

Installation and Use of the Calibrator

Upon completion, the calibrator should be connected to your receiver for test and alignment. The power requirements are 6.3 volts ac at 0.3 ampere for the filament, and 100 to 250 volts dc at 3 milliamperes for the plate. Remove your receiver from the cabinet and up-end it on your workbench. Find the audio amplifier tube (usually a 6V6-GT or a 6AQ5). The *filament* leads of the calibrator are soldered to pins 2 and 7 of the 6V6 socket, or to pins 3 and 4 of the 6AQ5 socket. You can consult a tube manual for filament connections if your receiver happens to use some other output tube. Next, solder the *B-minus* lead to a convenient grounding lug of a tube socket, or some other handy ground point. The *B-plus* lead should be soldered to the *screen* pin of the power output tube socket. This is pin 4 of the 6V6, or pin 6 of a 6AQ5 socket. Finally, attach the *output* lead of the calibrator unit to the antenna terminal of your receiver. Check your connections, and turn on the receiver. The filament of the calibrator tube should light. Turn on switch S1 of the unit and look for the 100 kc marker signals on the dial of your receiver. Turn off the calibrator and tune your receiver to station WWV or WWVH. If you have a local broadcast station operating on an even multiple of 100 kc (e.g.: 600 kc, 700 kc, etc.) you can tune to it. Wait until the tone is removed from WWV/WWVH, leaving only the time-tick signals. Turn on the calibrator and you will hear a low pitched audio note in the speaker of your receiver resulting from the heterodyne difference between the standard radio frequency signal and the harmonic of your calibrator. Using a fibre-blade screwdriver, slowly vary the setting of calibration control capacitor C1 until the heterodyne tone *drops in pitch* and stops. As you approach zero beat you will hear a very slow flutter between the two signals, which will grow slower as the frequencies coincide. Once you are tuned to zero beat, the calibrator is adjusted. The frequency calibration should be checked in this manner every few weeks or so to make sure that everything is in tip-top shape. If you wish, you can mount the calibrator inside the receiver cabinet. A small hole drilled in the corner of the receiver chassis will permit you to bring out the power leads to the calibrator unit.

Parts List for Crystal Calibrator

1—100 kilocycle precision crystal, X1. Peterson Radio Co. type Z-6A.
1—4-25 mmf variable ceramic capacitor, C1. Centralab 823-EN.
2—.01 mf, 600 volt ceramic capacitor, C2, C4. Centralab DD-1032.
1—220 mmf mica capacitor, C3. El-Menco CM15-E221-J.
1—30 mmf ceramic capacitor, C5. Centralab TCZ-30.
1—680,000 ohm, ½-watt resistor, R1. IRC type BTS or BW-½.
1—3,900 ohm, ½-watt resistor, R2.
1—100,000 ohm, ½-watt resistor, R3.
1—220,000 ohm, ½-watt resistor, R4.
1—10,000 ohm, 1-watt resistor, R5. IRC type BTA or BW-1.
1—Tube socket, 7-pin miniature, V1. Elco 300, with ground lugs.
1—Socket, crystal, 2-pin, X1. Millen 33102.
1—Switch, toggle, S1. Single pole-single throw. Carling S60A.
1—Phenolic terminal strip, T1. 4 lugs, plus two ground terminals. Cinch-Jones 54A
1—Chassis box, aluminum, 1½" x 2⅛" x 3½". LMB #000.
1—Tube, 6BA6, V1, RCA.
1—Grommet, ⅜-inch. General Cement 1042-E.
Misc. hardware: 4-40 nuts and bolts, 6-32 nuts and bolts. Soldering lugs, 10 feet of insulated hookup wire. Two ¼-inch metal spacers, General Cement 6765.

A DX Preselector For Your Receiver

You will be amazed at the way this inexpensive preselector will "hop-up" your receiver in the 14 mc to 30 mc range! Most receivers reveal a lack of gain and are bothered by image signals in the upper reaches of the shortwave spectrum. This DX preselector will remedy these troubles, providing you with over 10 decibels of signal gain plus high image attenuation. Best of all, this "signal sniffer" is inexpensive and easy to build! Enjoy improved shortwave reception when you attach this DX preselector to your present receiver! Have at it, men!

A *preselector* is an extra radio frequency amplifier stage placed between your receiver and the antenna. If your receiver has no r-f stage the addition of this sensitive unit will produce S9 signals where no signals were heard at all! The improvement in reception when the preselector is used with a receiver boasting an r-f stage is not quite so startling as this, but a signal boost of several S-units can immediately be noticed, particularly in the 20 mc to 30 mc region. The ten meter band which often seems lifeless will literally burst open with signals when a preselector is used. Troublesome image signals which hamper DX reception on the higher frequencies will

Fig. 7 An extra r-f stage will do wonders for your shortwave receiver. Shown above is an easy-to-build preselector covering the range of 14-30 mc. Single 6AS8 tube does double duty as the amplifier and rectifier. The r-f coils are contained within the aluminum shields, and the power transfer is at rear. "In-out" switch is at left of the chassis and a-c switch and gain control at the right.

fade into the background noise as the preselector is tuned for maximum strength of the wanted signal.

The Preselector Circuit

The schematic of the preselector is shown in Figure 8. A single 6AS8 dual-purpose miniature tube is employed as a pentode r-f amplifier and a power rectifier. The self-contained power supply permits the preselector to be used with any receiver, even the simple ac-dc types, since no power connections are made between the preselector and the receiver.

The grid and plate circuits of the 6AS8 r-f amplifier section cover the 14-30 mc range, tuned by a two-gang variable capacitor, C1A-B. No switching circuits are required to cover this range, making the task of construction and assembly of the unit a simple job. A switch S1A-B permits the preselector to be removed from the antenna circuit. An impressive demonstration of the DX qualities of the preselector may be heard by tuning your receiver to a weak station with the preselector removed from the circuit. When switch S1 is thrown to energize the unit, the signal will jump right up out of the background noise level as the main tuning control is adjusted.

The diode section of the 6AS8 serves as a half-wave rectifier, delivering approximately 120 volts dc to the pentode section of the tube. A simple resistance-capacitance filter network (R4-C7A-R5-C7B) insures hum-free operation. The signal gain is adjusted by potentiometer R3 in the cathode circuit of the r-f amplifier, preventing overload of the preselector by strong local signals. Sensitivity is maximum when R3 is turned fully clockwise. The ac-on switch S2 is manipulated by the control knob of potentiometer R3. Just turn it to the right and you're in business!

Your regular antenna is attached to the input terminals of the preselector, and a two-wire output line from the unit is attached to the antenna input terminals of your receiver. The main tuning dial (C1A-B) is adjusted for maximum strength as your receiver is tuned to a signal. Simple? You bet, and the preselector is a big help in finding those weak DX stations! Let's stop talking and get to work on it!

Preselector Construction

The parts layout and drilling template of the 3″ x 5″ x 7″ aluminum chassis are shown in Figure 9. The preselector is constructed upon one-half of a two-piece chassis box, the lower section of the box serving as a shield for the r-f circuitry. The following wiring and assembly instructions are presented in a step-by-step sequence to enable you to complete the project easily and correctly. Be sure to read each step all the way through before you start to do it. When the step is completed, check it off in the space provided. Do not solder any connections until told to do so. Let's get going!

( ) Lay out the chassis holes (Figure 9) to be drilled on the paper

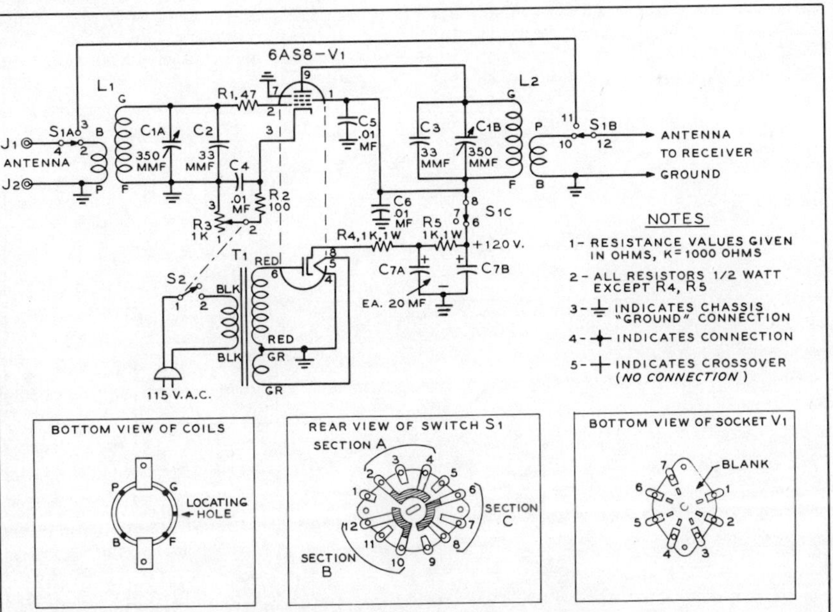

FIGURE 8—SCHEMATIC, PRESELECTOR

wrapper of the box. Drill the holes and remove the paper. Open the box and clean off the drill burrs with a file.

( ) Mount the 9-pin miniature socket V1 in hole A, using 4-40 hardware. Orient the socket so that pins 1 and 9 face the front of the chassis.

( ) Mount the two-gang tuning capacitor C1A-B beneath the chassis using three ¼-inch 6-32 screws. The shaft passes through hole B.

( ) Mount the coils L1 and L2 over the two 1¼-inch holes, using 6-32

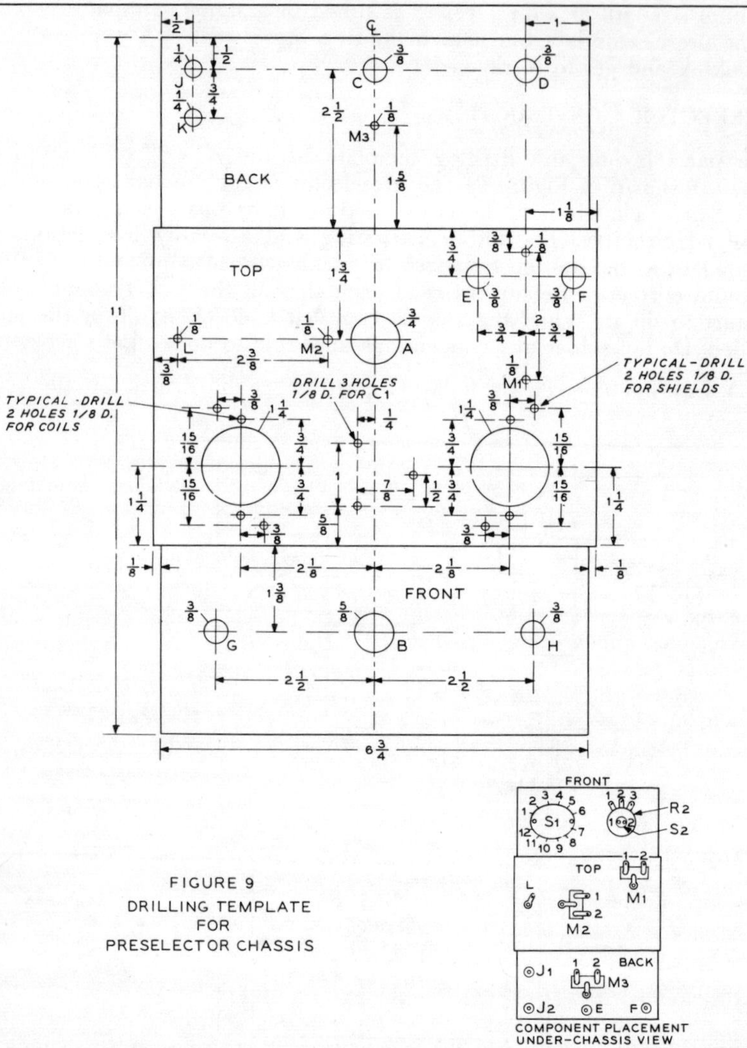

FIGURE 9
DRILLING TEMPLATE
FOR
PRESELECTOR CHASSIS

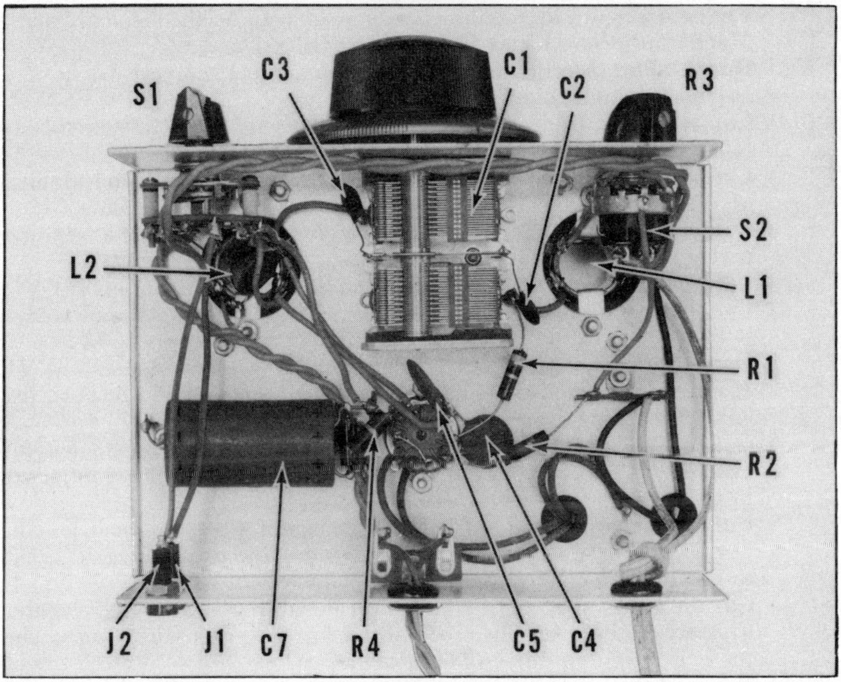

Fig. 10 Under-chassis view of preselector. Tuning capacitor C1A-B is at the center, with 6AS8 socket behind it. The output leads of unit pass through chassis hole at the rear of socket, and 115-volt cord is at the right. Antenna input jacks are at the left. The ground leads of padding capacitors C2 and C3 are soldered to the interstage shield of C1. The gain control potentiometer R3 is at the right.

hardware. The locating hole in the sides of both coils faces the tuning capacitor. Center each coil over its hole so that the terminals clear the chassis. Place two soldering lugs under the rear retaining nut of L2, and one lug under the front nut of L1.

( ) Place the aluminum shields over the coils, and fasten in position with 6-32 nuts.

( ) Mount the two-terminal phenolic strips M2 and M3 in position, using 6-32 hardware (Figure 9).

( ) Place 3/8-inch rubber grommets in chassis holes C, D, E, and F.

( ) Mount the power transformer T1 above the chassis using 6-32 hardware. Place phenolic terminal strip M1 under the front bolt of T1, and a ground lug under the rear bolt. Pass the transformer leads through grommets E and F.

( ) Mount rotary switch S1 in panel hole G, using lockwasher and nut supplied with switch. Adjust the stop of the switch to provide two positions. Orient the switch as shown in Figure 9.

( ) Mount potentiometer R3 in panel hole H, orienting as shown in Figure 9. If switch S2 is not attached to the potentiometer, it should be placed in position before R3 is mounted in the panel hole.

( ) Fasten a ground lug to the chassis at hole L, using 6-32 hardware.
( ) Mount tip jacks J1 and J2 in chassis holes J and K.
( ) Run a short length of bare wire from pin F of coil L1 to the adjacent ground lug (solder).
( ) Cut one black lead of transformer T1 to length and attach to terminal 2 of phenolic strip M1.
( ) Cut the other black lead of T1 to length and solder to terminal 2 of switch S2.
( ) Cut one green lead of T1 to length and solder to pin 5 of tube socket V1.
( ) Cut one red lead of T1 to length and solder to pin 6 of socket V1.
( ) Cut the remaining red and green leads of T1 to length and solder both of them to the grounding lug under the rear bolt of T1.
( ) Run a short lead from pin 4 of socket V1 (solder) to pin 7 of V1 (solder), and then to a socket grounding lug (solder). Be sure the wire clears pins 5 and 6.
( ) Cut the leads of a .01 mf ceramic capacitor C4 to ½-inch length. Connect C4 between pin 3 of socket V1 and the ground lug adjacent to pin 1.
( ) Cut the leads of a .01 mf ceramic capacitor C5 to ½-inch length. Connect C5 between pin 1 of socket V1 and the ground lug adjacent to pin 1 (solder).
( ) Cut the leads of a .01 mf ceramic capacitor C6 to ½-inch length. Connect C6 between pin F of coil L2 and one grounding lug under the coil retaining nut (solder).
( ) Run a short length of bare wire from terminal B of coil L2 to the adjacent ground lug (solder).
( ) Cut the leads of a 100 ohm resistor R2 (brown-black-brown) to 1-inch length. Connect between pin 3 of socket V1 (solder) and terminal 1 of phenolic strip M1.
( ) Cut the leads of a 47 ohm resistor R1 (yellow-purple-black) to 1-inch length. Connect R1 between pin 2 of tube socket V1 (solder) and the *rear right stator* terminal of tuning capacitor C1A.
( ) Run a lead from pin G of coil L1 (solder) to the rear stator terminal of capacitor C1A.
( ) Cut the leads of a 33 mmf ceramic capacitor C2 to ¾-inch length. Connect C2 between the rear right stator terminal of C1A (solder) and the top hole in the interstage shield of capacitor C1A (solder).
( ) Cut the leads of filter capacitor C7A-B to ⅝-inch length. Orient the ground (–) lead of C7 towards soldering lug L and the two positive leads towards phenolic strip M2. Bend the positive leads of C7 so that each lead passes through a hole in each terminal of M2. Twist the ground lead around lug L (solder). Dress the capacitor against the chassis.
( ) Cut the leads of a 1K, 1-watt resistor R5 (brown-black-red) to ⅜-inch. Connect R5 between terminals 1 and 2 of strip M2.
( ) Cut the leads of a 1K, 1-watt resistor R4 (brown-black-red) to ½-inch. Connect R4 between pin 8 of tube socket V1 (solder) and terminal 2 of strip M2 (solder).
( ) Run a lead from terminal 1 of strip M2 (solder) to terminal 6 of

switch S1, section C (solder).

( ) Run a lead from pin 1 of socket V1 (solder) to pin F of coil L2.

( ) Run a lead from pin 9 of socket V1 (solder) to pin G of coil L2.

( ) Run a lead from pin F of coil L2 (solder) to terminal 8 of switch S1, section C (solder).

( ) Run a lead from terminal 11 of switch S1, section B (solder) to terminal 3 of S1, section A (solder). Make sure the wire clears the other switch contacts.

( ) Run a lead from pin G of coil L2 (solder) to the *front stator* terminal of tuning capacitor C1B.

( ) Clip the leads of a 33 mmf capacitor C3 to a length of ¾-inch. Connect C3 between the front stator terminal of C1A (solder) and the top hole in the interstage shield of capacitor C1 (solder).

( ) Connect a lead between pin B of coil L2 (solder) and antenna terminal jack J2 (solder).

( ) Connect a lead between terminal 4 of switch S1, section A (solder) and antenna terminal jack J1 (solder).

( ) Cut two lengths of wire eight inches long. Strip and tin one end of each wire. Twist the wires together. Solder one lead to pin B of coil L2. Solder the other lead to terminal 12 of switch S1, section B.

( ) Pass the twisted leads between capacitor C7 and terminal strip M2, up to terminal strip M3. Cut the wires to length and solder one wire to terminal 1, and the other wire to terminal 2 of M3.

( ) Run a lead from terminal 1 of strip M1 (solder) to terminal 2 of potentiometer R3 (solder).

( ) Run a lead from pin P of coil L1 (solder) to terminal 3 of potentiometer R3 (solder).

( ) Cut two lengths of hookup wire one foot long each. Strip and tin one end of each wire. By ink or other means, mark the insulation at both ends of *one wire* for identification. Twist the wires together.

( ) Solder the *identified* lead to pin B of coil L2.

( ) Solder the *unidentified* lead to terminal 2, switch S1, section A.

( ) Dress the twisted wires across the top of tuning capacitor C1, close to the front panel (Figure 10). Pass the leads around potentiometer R3 and downwards towards coil L1.

( ) Trim the leads to length. Solder the *identified* lead to pin P of coil L1.

( ) Solder the *unidentified* lead to pin B of coil L1.

( ) Pass the 115-volt line cord through grommet D from the rear of the chassis. Allow about six inches of the cord to project inside the chassis. Tie a knot in the cord adjacent to the grommet.

( ) Trim one lead of the cord and solder to terminal 2 of strip M1.

( ) Trim the other lead and solder to terminal 1 of switch S2.

( ) Cut two lengths of hookup wire two feet long. Strip and tin both ends of the wire. By ink or other means, mark the insulation at both ends of one wire for identification. This is the *antenna lead.* Twist the leads and pass them through grommet C on the rear of the chassis.

( ) Solder the *antenna lead* to terminal 1 of strip M3.

( ) Solder the *other lead* to terminal 2 of M3.

( ) Adjust the two variable mica padding capacitors on the side of the main tuning capacitor so that the spring plate is compressed flat against the mica sheet.

( ) Inspect the wiring for accidental shorts, unsoldered joints and grounds. Check your wiring against Figure 8.

( ) Assemble the chassis box using the 6-32 screws supplied with the box.

( ) Insert the 6AS8 tube in the socket. Place the knobs on S1, C1, and R3. When S1 is turned counter-clockwise, the preselector is out of the circuit. Clockwise rotation of S1 connects the preselector between your antenna and the input terminals of the receiver.

Testing and Using the Preselector

The preselector is placed near your receiver and your antenna connected to jacks J1 and J2. If a single wire antenna is in use, it should be connected to J1 and an external ground connected to J2. The twisted leads from the converter connect to the antenna input terminals of your receiver. The *antenna* lead is attached to the "hot" antenna terminal, and the *unmarked* lead connects to the ground terminal, as shown in Figure 8. *Note!* If your receiver is an ac-dc set it is important that no dc connection exists between the chassis of the preselector and that of the receiver in order to reduce the shock hazard. Most ac-dc sets have an isolating capacitor between the "ground" terminal and the chassis of the receiver. If this capacitor is not included in the set, or if it cannot be found, it is wise to insert a .01 mf, 600 volt ceramic capacitor in series with *each lead* from the preselector to the antenna and ground terminals of the receiver. This will in no way affect the operation of the preselector, yet will remove the danger of accidental shock.

When the preselector is turned on, and S1 turned clockwise the unit is in operation. The filament of the 6AS8 tube should light, and a voltage of about 120 may be measured between the positive terminal of C7B and ground if a high resistance dc voltmeter is at hand. Tune the receiver to some frequency in the 14-30 mc range and advance potentiometer R3 fully clockwise. Slowly tune the preselector dial and you will note the signal suddenly jumps upward in intensity. Tuning is quite sharp, so take it easy! The gain of the preselector may be reduced by backing off on R3. A few minutes spent tuning the receiver and preselector together will quickly show the optimum setting of R3. If you doubt the efficiency of the preselector, just throw S1 to the left and watch the signal strength drop on the S-meter of your receiver! You will find this little unit an invaluable aid to better DX reception!

The tuning range of the preselector may be moved about by playing with capacitors C2 and C3. If these capacitors are removed from the unit, the top of the tuning range will be approximately 36 mc, and the lower tuning limit will rise to about 16 mc. If C2 and C3 are increased in value to 56 mmf each, the tuning range will be lowered to about 10 mc, but the upper tuning limit will not include the 10 meter amateur band. A final word: The bottom part of the chassis-box forms a shield around the tuning circuits. If you leave it off, the preselector may tend to oscillate. Make sure its on tight!

Parts List for Preselector

1—Dual section variable capacitor, 350 mmf, C1A-B. J. W. Miller 2112.
2—33 mmf ceramic capacitor, C2, C3. Centralab MD-330.
3—.01 mf, 600 volt ceramic capacitor, C4, C5, C6. Centralab DD-1032.
1—Dual filter capacitor, 20-20 mf at 250 volts. Sprague TVA 2515.
2—Coils, r-f, L1, L2. J. W. Miller E-727A.
2—Coil shields, J. W. Miller S-727.
1—47 ohm, ½-watt resistor, R1, IRC type BTS or BW-½.
1—100 ohm, ½-watt resistor, R2.
1—1,000 ohm potentiometer, R3, with switch S2. Centralab B5-S.
2—1,000 ohm, 1-watt resistor, R4, R5. IRC type BTA or BW-1
1—Tube socket, 9-pin miniature, V1. Elco 370 with ground lugs.
1—Transformer, 125 volts at 15 ma., 6.3 volts, T1. Chicago-Standard PS-8415.
1—Switch, rotary, 3-pole, 3-position, S1. Centralab PA-1007.
3—Phenolic terminal strip, M1, M2, M3. 2 lugs. Cinch-Jones 52.
1—Tube, 6AS8, V1. RCA
1—Chassis box, aluminum, 7" x 5" x 3". LMB 145
2—Nylon tip jack and plug, J1, J2. Johnson 105-602 and 105-302.
1—Line cord and plug. Belden 1765B.
4—Grommet, ⅜-inch. General Cement 1042-E.
2—Knob. General Cement 1170.
1—Dial. National type AM.
Misc. hardware: 4-40 nuts and bolts, 6-32 nuts and bolts. Soldering lugs, 15 feet of insulated hookup wire.

Greater Selectivity With A Home-built Q-Multiplier

Troubled with QRM? Are the stations piled upon themselves on the dial of your receiver? Additional selectivity is what you need! No, you don't have to buy a new receiver—just add a Q-Multiplier to your present set. The Q-Multiplier is a regenerative circuit that will boost the selectivity of your receiver and reduce the effects of adjacent channel interference. You can build this simple, effective unit for less than seven dollars! If your receiver has an i-f of 455 kc, you can enjoy improved reception with this Q-Multiplier!

Congested shortwave bands demand the use of some form of selectivity control for your receiver. A crystal filter, a mechanical filter, or a Q-Multiplier should be used for best reception. Of these three devices, the Q-Multiplier is by far the least expensive to build. Its operating characteristics are excellent, and you will find that it will add razor-sharp selectivity to your receiver—just the ticket for peeling off those top layers of QRM!

The Q-Multiplier is a one tube amplifier that is connected to the i-f amplifier of your receiver. It is a positive feedback device having a highly selective tuned circuit that acts as a narrow gate through which the i-f signals of your receiver must pass. You can open or close this gate by adjusting the controls of the Q-multiplier. At maximum selectivity, only one signal is allowed through the selectivity gate. No circuit changes need be made to your receiver in order to use the Q-multiplier. The small amount of power required by the unit may be "stolen" from any receiver that is not of the ac-dc type.

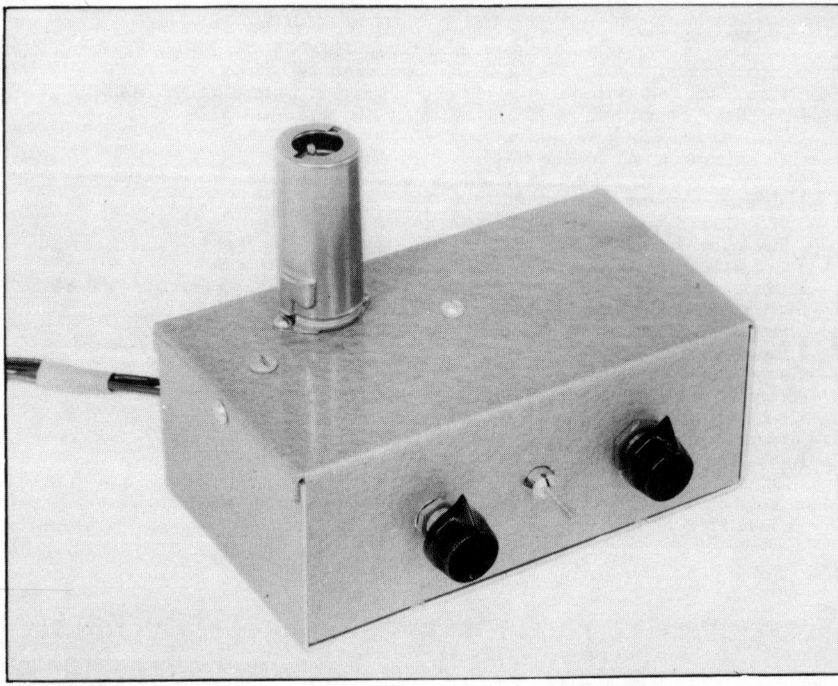

Fig. 11 Selectivity of receiver may be improved by addition of this easy-to-build Q-Multiplier. The tuning control C2 is at left, with gain control R4 at the right.

The Q-Multiplier Circuit

The schematic of the Q-Multiplier is shown in Figure 12. A single miniature 6AV6 triode is used as the multiplier tube with a ferrite "loopstick"-type coil connected between the grid and the plate. The Q-Multiplier is tuned to the intermediate frequency (455 kc) of the receiver by adjusting the slug of coil L1, and signal peaking is accomplished by the tuning capacitor C2. The gain of the unit is controlled by potentiometer R4 in the cathode circuit of the 6AV6 tube. One side of the highly selective tuned circuit (L1-C2-C3-C4) is coupled through an isolating capacitor C1 to the i-f amplifier of your receiver.

Q-Multiplier Construction

The Q-Multiplier is compactly built on one-half of a two-piece aluminum chassis box measuring 2" x 3" x 5" in size. The parts layout and drilling template are shown in Figure 13. The unit is small enough to sit atop your receiver, permitting very short connections to be made between the Q-Multiplier and the "innards" of your set. The following wiring and assembly instructions are presented in a step-by-step sequence to enable you to complete the project easily and correctly. Be sure to read each step all the way through before you start to do it. When the step is completed, check it off

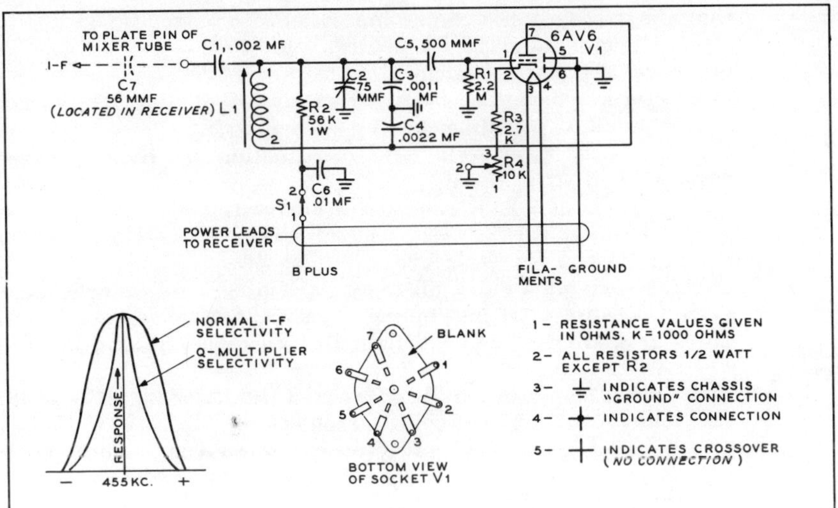

FIGURE 12—SCHEMATIC, Q-MULTIPLIER

in the space provided. Do not solder any connections until told to do so.

( ) Lay out the chassis holes (Figure 13) to be drilled on the paper wrapper of the box. Drill the holes and remove the paper. Open the box and clean off the drill burrs with a file.

( ) Mount the 7-pin miniature tube socket V1 in hole H, using 4-40 hardware. Orient the socket so that pins 1 and 7 face the right side of the chassis when viewed from beneath. Place a phenolic terminal strip M1 beneath the nut at the right side of the socket.

( ) Mount a grounding lug (soldering lug) in hole A using 6-32 hardware.

( ) Mount a phenolic terminal strip M2 in chassis hole B.

( ) Mount a phenolic terminal strip M3 in chassis hole J.

( ) Place 3/8-inch rubber grommets in chassis holes C and D.

( ) Mount the 75 mmf tuning capacitor C2 in hole E, using the collar nut supplied with the unit. The *rotor lug* of the capacitor faces upwards towards you.

( ) Mount the ferrite loopstick coil L1 in chassis hole F. Notice that the slug end of the coil has small metal "fingers." Push the coil through the chassis hole from the inside until the metal fingers pass through the chassis and snap over the edge of the hole. A firm, easy pressure will do the job. Rotate the coil until the two terminals are oriented as shown in Figure 14.

( ) Mount the 10K potentiometer R4 in chassis hole G, using the hardware supplied with the unit. If switch S1 is not attached to the potentiometer, fasten it in position before you mount R4.

( ) Run a short length of bare wire from pin 5 of socket V1 (solder) to pin 6 of V1 (solder) to the grounded terminal 1 of strip M1.

( ) Clip the leads of a 2.2-megohm resistor R1 (red-red-green) to ½-inch length. Connect R1 between pin 1 of socket V1 and the grounded terminal 1 of strip M1.

( ) Connect a lead from the grounded terminal 1 of strip M1 (solder) to terminal 2 of potentiometer R4 (solder).

( ) Connect a lead from terminal 3 of potentiometer R4 (solder) to terminal 2 of strip M1.

( ) Clip the leads of a 2.7K resistor R3 (red-purple-purple) to ½-inch length. Connect R3 between pin 2 of socket V1 (solder) and terminal 2 of strip M1 (solder).

( ) Clip the leads of a .0011 mf mica capacitor C3 to ⅝-inch length. Place C3 between L1 and tuning capacitor C2. Solder one lead of C3 to grounding lug A, and twist the other lead around lug 1 of coil L1.

( ) Run a short length of bare wire between the *stator* lug of capacitor C2 (solder) and lug 1 of coil L1 (solder).

( ) Clip the leads of a .0022 mf mica capacitor C4 to ⅝-inch length. Place C4 between coil L1 and potentiometer R4. Solder one lead of C4 to grounding lug A, and twist the other lead around lug 2 of coil L1.

( ) Run a lead between lug 2 of coil L1 (solder) and pin 7 (solder) of tube socket V1.

( ) Clip the leads of a 500 mf mica capacitor C5 to ¾-inch length. Solder one lead of C5 to pin 1 of socket V1. Attach the other lead to the tip of one *stator* rod of capacitor C2.

( ) Clip the leads of a .002 mf ceramic capacitor C1 to ¾-inch length. Solder one lead of C1 to the tip of the stator rod of C2, and attach

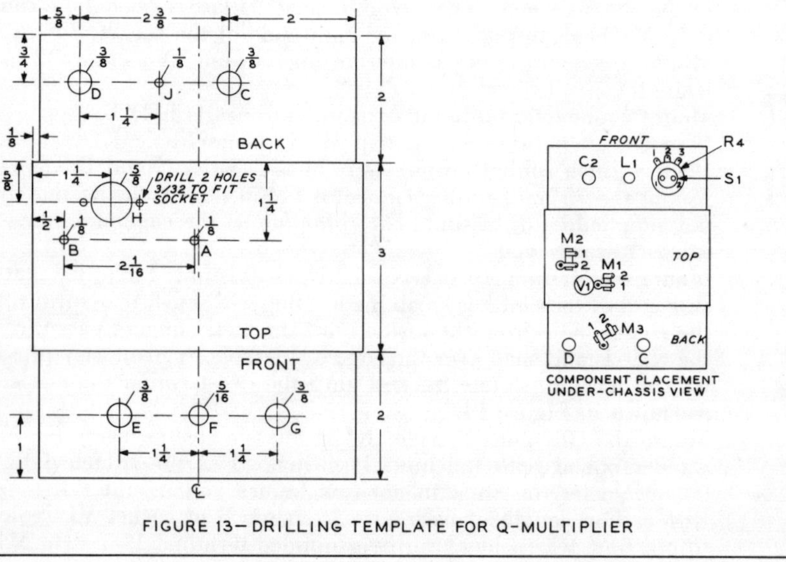

FIGURE 13—DRILLING TEMPLATE FOR Q-MULTIPLIER

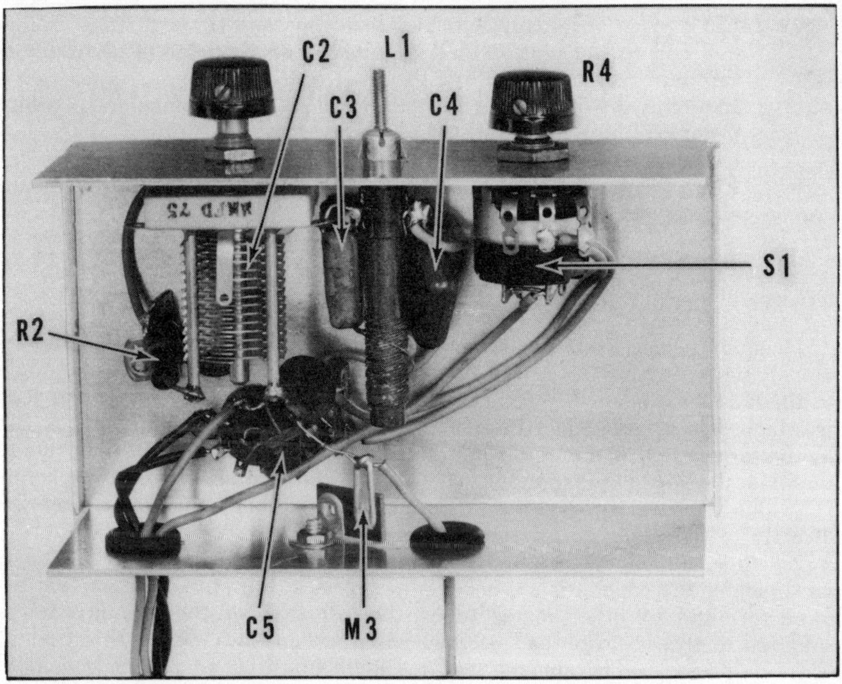

Fig. 14 Under-chassis view of Q-Multiplier showing the placement of major components. The output coupling capacitor C1 is mounted between the stator (fixed) terminal of tuning capacitor C2 and terminal strip M3 on rear of chassis. The bypass capacitor C6 is mounted across terminals of phenolic strip M2, just visible behind the resistor R2. The Q-Multiplier is resonated to the intermediate frequency of receiver by tuning slug of L1. Fine tuning is done with capacitor C2.

the other lead to the ungrounded terminal 2 of strip M3.

( ) Clip the leads of a .01 mf ceramic capacitor C6 to ½-inch length. Connect C6 between terminal 1 and 2 (solder) of strip M2.

( ) Run a lead from terminal 1 of strip M2 to terminal 1 of switch S1 (solder).

( ) Clip the leads of a 56K, 1-watt resistor R2 (green-blue-orange) to ⅝-inch. Connect R2 between terminal 1 of strip M2 (solder) and the free stator rod of capacitor C2 (solder).

( ) Cut five pieces of hookup wire each two feet long. Strip and tin both ends of all wires.

( ) Twist two of these wires together their full length. Solder the end of one wire to pin 3 of socket V1. Solder one end of the other wire to pin 4 of V1. Mark the free end of these leads with a piece of adhesive tape labelled *filaments*. Pass the leads through grommet D.

( ) Solder a third lead to pin 5 of socket V1. Mark the free end *ground*. Pass the lead through grommet D.

( ) Solder the last lead to terminal 2 of switch S1. Mark the free end *B-plus*. Pass the lead through grommet D.

( ) Cut a piece of hookup wire two feet long, and tin both ends. Solder one end to the ungrounded terminal 2 of strip M3. Pass the lead through grommet C. Mark the lead *i-f*.

( ) Examine all your wiring for accidental shorts and unsoldered joints. Compare the wiring with Figure 12. Attach the bottom of the box to the chassis using the 6-32 screws supplied with the box.

( ) Place the tuning knobs on C2 and R4. Place the 6AV6 tube in the socket, and the shield over it.

Testing the Q-Multiplier

The four power leads are attached to your receiver in the manner previously described for the crystal calibrator. The twisted *filament* leads attach to the filament pins of the audio amplifier tube socket, and the *ground* lead is attached to a convenient ground point in the set. The *B-plus* lead is run to the screen pin of the audio amplifier tube socket. The *i-f* lead is run to the plate pin of the mixer tube socket through a 56 mmf coupling capacitor C7 which is placed *at the mixer socket.* One lead of the capacitor is soldered directly to the plate pin of the tube socket. If your receiver uses a 6SA7 mixer tube, attach one lead of the capacitor to pin 3. If a 6BE6 tube is used, solder the capacitor to socket pin 5. Check your tube manual for the plate terminal of other mixer tubes. The *i-f* lead of the Q-multiplier is soldered to the opposite lead of this coupling capacitor. Make this lead as short as possible. The slickest way to do this task is to mount a coaxial receptacle J1 on the back of your receiver chassis and connect the lead from the coupling capacitor C7 to the center pin of the coaxial receptacle. Place a matching plug P1 on the *i-f* lead from the Q-Multiplier and plug it in the receptacle. The other leads to the "innards" of the receiver which obtain power for the Q-Multiplier may be passed through a small hole drilled in the corner of the receiver chassis. Before you get carried away and drill this hole for the power leads, look to see if your receiver has an auxiliary power receptacle on the back of the chassis! If one is there, you can make your power connections to it, and save yourself some extra work.

When the Q-Multiplier is properly connected to the receiver, turn it on, and turn on your receiver. The 6AV6 tube should light. Rotate potentiometer R4 fully clockwise, and set capacitor C2 half-meshed. Tune in a local broadcast station and watch the S-meter of your receiver as you turn the slug of coil L1 with the blade of a small screwdriver. When the slug is about ½-inch out of the form, the S-meter of your receiver should kick up, and you will hear a howl in the speaker as the Q-Multiplier breaks into oscillation. Back off on potentiometer R4 until the howl stops, and readjust the slug of coil L1 for maximum S-meter reading. Tuning capacitor C2 should peak the reading, and also drop it to a low value when C2 is tuned off-resonance. The point of maximum selectivity is reached just before the Q-Multiplier breaks into oscillation. Potentiometer R4 will have to be retarded from this point for best phone reception.

Once the slug of coil L1 has been adjusted, it may be left alone and all tuning done with capacitor C2. The signal is peaked with C2, and control R4 advanced until the desired degree of selectivity is obtained. As a final step, the primary and secondary trimmers of the first i-f transformer in your receiver should be repeaked as the Q-Multiplier tends to detune this stage a bit.

As a check on the effectiveness of the Q-Multiplier, adjust it for maximum selectivity, as indicated by a ringing sound in the speaker. Then pull the plug out of the rear jack of the receiver and notice the difference!

When you are using the Q-Multiplier for phone reception, retard R4 for best fidelity of the signal. If QRM appears, advance R4 until the best balance is obtained between fidelity and interference. Rock the tuning dial of your receiver back and forth until the point of best reception is found.

The Q-Multiplier may be allowed to break into oscillation for c-w reception. The b-f-o of the receiver is not used in this case, as the Q-Multiplier provides the necessary beat note for c-w reception. Extreme selectivity can be achieved with this form of reception. It is not recommended for single-sideband phone reception, however, as the selectivity is too sharp for speech reproduction.

Parts List for Q-Multiplier

1—.002 mf ceramic capacitor, C1. Centralab DD-202.
1—75 mmf variable capacitor, C2. Bud LC-1645.
1—.0011 mf mica capacitor, C3. El-Menco CM30-112.
1—.0022 mf mica capacitor, C4. El-Menco CM30-222.
1—500 mmf mica capacitor, C5. El-Menco CM15E-501J.
1—.01 mf ceramic capacitor, C6. Centralab DD-1032.
1—56 mmf ceramic capacitor, C7. Centralab MD-560.
1—Coil, "loopstick" ferrite, L1. J. W. Miller 6300.
1—2.2 megohm, ½-watt resistor, R1. IRC type BTS or BW-½.
1—56,000 ohm, 1-watt resistor, R2. IRC type BTA or BW1.
1—2,700 ohm, ½-watt resistor, R3.
1—10,000 ohm potentiometer, R4 with switch S1. Centralab B-14S.
1—Tube Socket, 7-pin miniature, with shield, V1. Elco ST-234, with 126 shield.
2—Phenolic terminal strip, M1, M2. 1 lug. Cinch-Jones 51.
1—Tube, 6AV6, V1. RCA.
1—Chassis box, aluminum, 5¼" x 3" x 2". LMB 136.
2—Grommet, ⅜-inch. General Cement 1042-E.
2—Knob. General Cement 1118.
Misc. hardware: 4-40 nuts and bolts, 6-32 nuts and bolts. Soldering lugs, 15 feet of insulated hookup wire.

CHAPTER VII

Better Antennas for Your Receiver

A receiver is no better than its antenna! Some listeners buy an expensive receiver, hang a hunk of wire on it, and are disappointed at the results they obtain. "Aw, the receiver is no good. Shortwave reception is the bunk," they say. This situation can be avoided if a little care and thought is put into the erection of the receiving antenna. Big dividends in the form of improved reception may be gained with the use of an efficient, well planned antenna, such as one of the various types described in this chapter. Give your receiver and yourself a break! Put up a good antenna for best DX reception!

An *antenna* (or *aerial)* is a device that will capture electromagnetic energy from space and convert it to a movement or drift of electrons. The most common form of antenna is a piece of copper wire strung between two supports. The efficiency of such a simple device is a function of the length of the wire as compared to the length of the radio wave, and the angle at which the radio wave strikes the wire. Every type of radio antenna has directional characteristics causing it to intercept energy better in certain directions than in others.

One end of the antenna wire may be brought into the house and connected directly to the antenna terminal of the receiver, or the antenna may be separated from the receiver and connected to it by a "radio hose" called a *transmission line*, which conducts the energy from the antenna to the receiver with a minimum of loss.

What Is A Good Antenna?

A good antenna is the key to successful shortwave reception. Time and effort spent in making the best antenna installation possible will result in greatly improved reception. No matter what form or type of antenna you use, it must follow some rather fundamental rules, or it will not be worth the time and money required to put it up. Here are these rules:

1—The antenna should be erected as high in the air as possible. Twenty

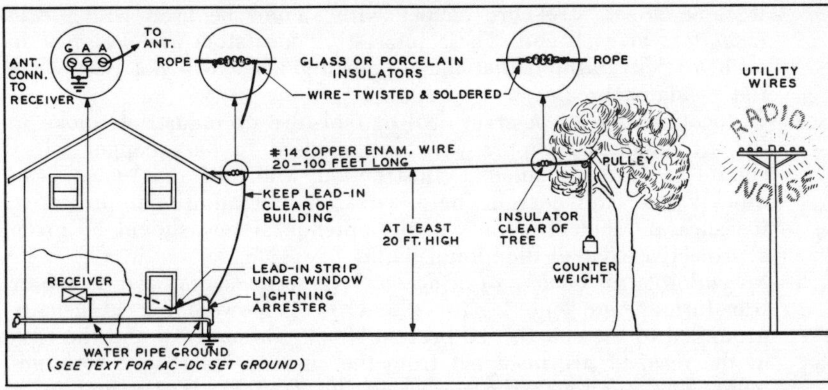

Fig. 1 The single wire antenna shown above is simple and inexpensive to build and will produce good results over the whole shortwave spectrum. The antenna should be erected as high as possible, and at right angles to any nearby power lines. Make sure the lead-in does not touch the gutter pipes.

feet above the ground may be considered to be a minimum height. Some amateur transmitting and receiving antennas are placed eighty or ninety feet in the air to obtain optimum results! Get your antenna as high in the air as you can.

2—The location of the antenna with respect to nearby objects is very important. Telephone lines and power lines exhibit a shielding effect to a nearby antenna. Stucco buildings having wire mesh in the walls are also a detriment to best reception. If possible, place your antenna so that its length is at right angles to nearby utility wires. Keep your antenna clear of the walls of nearby buildings, or try and raise the antenna over the top of the building.

Many power lines, particularly older ones carrying high voltage, have poor hardware and leaky insulators which produce severe electrical interference. This will show up as a loud buzzing noise covering most of the shortwave spectrum. The farther the antenna is from the power line, the smaller will be the amount of noise pickup. Unfortunately, high voltage transmission lines can emit a loud radio racket that can be heard on a sensitive receiver a mile or two removed from the line! Orientation of the antenna for minimum noise pickup is a partial solution in such a case.

3—The antenna installation should be sturdy and well made. Occasionally you will read in the paper of a radio antenna falling across a high tension line and electrocuting the user of the receiver! Makeshift connections and haywire are definitely "out" when you erect an antenna! Stay away from power lines, and make sure that if your antenna should happen to come down, it will not hit a utility cable! If the antenna is attached to a tree, some allowance should be made for the sway of the tree in a wind. A rope, pulley, and weight should be used at the tree end of the antenna as shown in Figure 1. All joints in the antenna wires should be well soldered. In windy locations, or if the antenna might fall into the street if it came down, steel-core copper wire should be used with *strain insulators* at each end. These interesting insulators are designed in such a tricky fashion that the supporting wires will not part when the insulator breaks.

4—In locations having a great deal of moisture or industrial smoke in the air, enameled copper wire (as opposed to bare copper wire) should be used for antenna construction. Moisture and "smog" can quickly rust and corrode bare wires. As an additional protective measure, all soldered joints in the antenna system should be given a protective coat of aluminum paint.

5—*Special note for owners of ac-dc sets!* These receivers have no power transformer and one leg of the 115-volt power line is usually grounded to the chassis. To prevent shock, all controls, and the case of the receiver are insulated from the chassis. Make sure your antenna does not touch the chassis of the receiver. If you employ a ground connection, make sure the "ground" terminal of the set is insulated from the chassis, and that the ground connection is made through a capacitor within the set. If such a capacitor can not be found, make your ground connection to the receiver through a 0.01 mfd, 600 volt paper capacitor.

Armed with these five basic rules, let's examine some transmission lines, and some typical antennas that are easy to erect and inexpensive.

Antenna Resonance

For any antenna there is one frequency, called the *resonant frequency,* at which the capacitive and inductive reactances are exactly equal and neutralize each other, leaving only the radiation resistance of the antenna to oppose the flow of current in the circuit. The resonant frequency of a simple antenna is a function of its physical length. If the antenna is not the correct length for resonance at the desired frequency of operation, it can accordingly be brought into resonance by adding series inductance to equalize the capacitive reactance of an antenna that is too short, or by adding series capacitance to neutralize the inductive reactance of an antenna that is too long.

Energy Transfer to the Antenna

A resonant antenna wire suspended well in the air makes a fine antenna, but it does little good unless some means are provided to conduct the radio energy from the antenna to the receiver, or from the transmitter to the antenna. To accomplish this, a radio "hose" called a *transmission line,* or *feeder,* is used to pipe the energy back and forth between the radio equipment and the antenna. Four common types of transmission line are shown in Figure 2. Although there are physical differences between these lines, they all have the common property called the *characteristic impedance* of the line. This impedance is expressed in ohms, and is determined by the physical and electrical properties of the line, such as the size of the wire, the spacing between the conductors, and the type and amount of insulating material used. Two common impedance values of line are 52 ohms and 300 ohms. Many types of inexpensive television transmission line are available

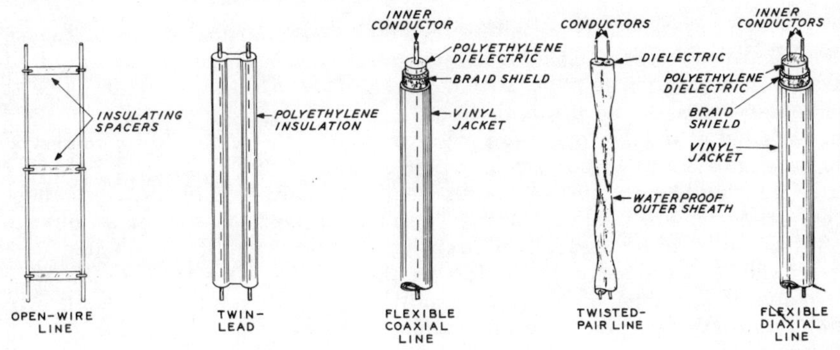

Shown above are various types of balanced and unbalanced transmission lines suitable for high frequency use. The most popular types are the TV-style 300-ohm twin lead and the 52-ohm coaxial unbalanced line. The transmission line may be thought of as a "radio hose," piping the energy extracted from the radio wave by the antenna down to the radio equipment. An example of the use of a coaxial transmission line is shown in Figure 6.

in these values, and it would be a smart move to design the antenna so that it can efficiently employ these popular line impedance values.

An impedance notation of "52 ohms" or "300 ohms" does not mean that the transmission line has a resistance equal to that value. In fact, efficient lines made of copper wire have resistance values less than one ohm or so. The characteristic impedance notation of "so-many-ohms" means that the line is designed to be used with an antenna that presents that value of *terminating resistance* or *radiation resistance* to the transmission line. To clarify these last two terms, let's look at the antenna once again.

Radiation Resistance

The radiation resistance of a dipole—or any other antenna for that matter—is expressed in ohms and may be defined as that value of resistance which, when substituted for the antenna, will dissipate in the form of heat the same amount of power as is radiated into space by the antenna. The actual value of radiation resistance is determined by the length and size of the antenna compared to the radio wave, and the proximity and character of objects located near the antenna. The radiation resistance of any antenna may be measured with appropriate instruments, and usually runs in the neighborhood of 30 to 300 ohms for simple resonant antennas usually employed by Novices and Technicians. The radiation resistance is measured at that portion of the antenna having the greatest value of current. In the case of the dipole, this is the center point of the antenna.

The Single Wire Antenna

For general shortwave reception the single wire antenna shown in Figure 1 is a hard one to beat. The *flat top* portion should be twenty to one hundred feet long for best results on all bands. The *lead-in* portion of the antenna should be as short as possible. Good, four inch glass insulators are used at each end of the flat top, and a piece of insulated wire should be employed to pass the antenna through the window or wall of the building. A lightning arrestor should be connected to the lead-in if you live in an area where thunderstorms are likely to occur. The antenna must be free of all objects and should be in the clear. Tree branches scraping against the antenna wire are likely to produce weird noises in the receiver. A simple single wire antenna of this type is relatively nondirectional, although best results will be had broadside to the wire rather than along its length. Thus, for general reception in an east-west direction, the wire should run in a north-south direction.

Most shortwave receivers have two antenna terminals and a single ground terminal. The lead-in wire should be attached to one of the antenna terminals, and the other antenna terminal should be connected to the ground terminal of the set with a short length of wire. Try reversing the connections to the antenna terminals to see which arrangement will provide best reception. In some instances reception will be improved and noise will be decreased by running a wire from the ground terminal of the receiver to a nearby ground connection. A four foot pipe may be driven into the ground for such a termination, or the ground wire may be attached to a cold water pipe in the house. A gas pipe should not be used for a ground connection.

The T-Antenna And Its Cousin, The Windom

In some instances it might be more handy to have the lead-in wire near the center of the antenna, rather than at one end (Figure 3). An antenna

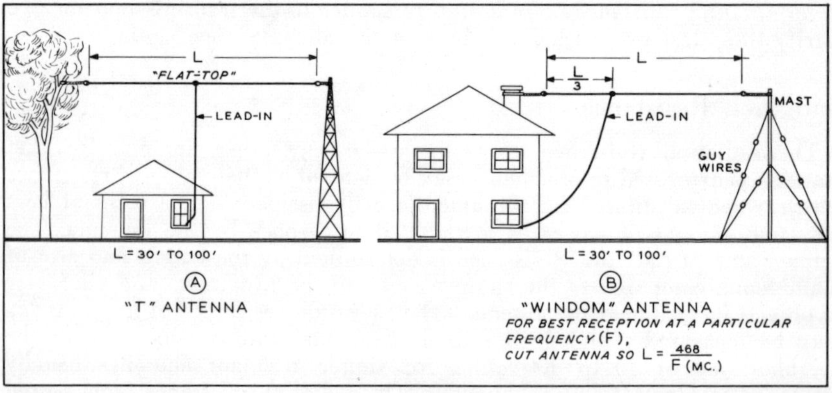

Fig. 3 A single wire lead-in may be used with your shortwave antenna. Vertical section of lead-in is particularly susceptible to pickup of man-made noise.

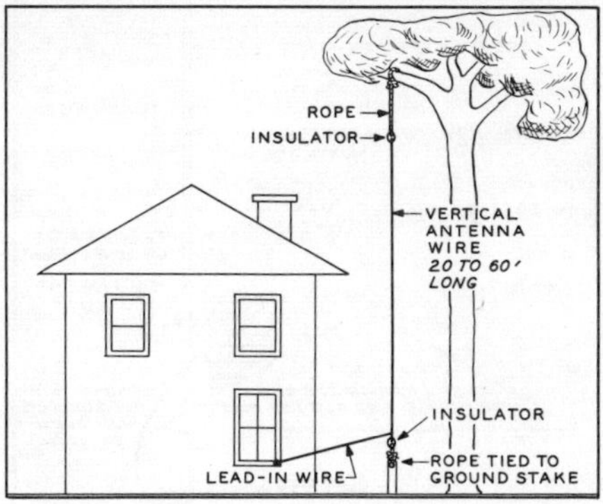

Fig. 4 Simple vertical antenna may be supported at top end by nearby tree. Installation should be protected with lightning arrestor mounted at window sill of the house. Antenna should be kept well clear of any building.

of this type is called a *T-antenna.* The horizontal portion of the antenna should be thirty to ninety feet long, and the vertical portion can be any length. The lead-in wire drops vertically from the center of the flat top. An unbalanced version of the T-antenna is the *Windom,* shown in Figure 3B. The Windom lead-in is tapped on the flat top approximately one-third the distance from one end to provide a good impedance match between the flat top and the lead-in wire. Either type of antenna will work well at any frequency, but the Windom antenna is actually tuned to a specific frequency which is a function of L. When used with the simple antenna tuner described later in this chapter it will prove to be one of the best types of antennas intended for general shortwave reception. The Windom antenna may be cut for peak reception in one of the amateur bands, or one of the shortwave broadcast bands with the assurance that good reception may still be enjoyed at other frequencies, even if the antenna tuner is not used.

The Vertical Antenna

The *vertical antenna* is shown in Figure 4. It is relatively non-directional. (As some wit once said, "The vertical antenna receives equally poorly in all directions.") This is a harsh judgment, as the vertical antenna is an excellent performer. Its main disadvantage is that it is very responsive to atmospheric and man-made noise—much more so than a horizontal antenna.

The obvious advantage of the vertical antenna is that it takes up very little ground space. It can be bolted to a chimney, or attached to the side of a house. If you live in an area free of severe automobile ignition noise and your static level is reasonably low, you might think about using some sort of vertical antenna, such as the one shown in Figure 5. A twenty or thirty foot vertical antenna will work well over the entire shortwave spectrum. The over-all length of antenna plus lead-in should not exceed about thirty feet, or antenna operation at the higher frequencies will be impaired.

The Vertical Ground Plane Antenna

The *ground plane* antenna is a close relative to the vertical antenna.

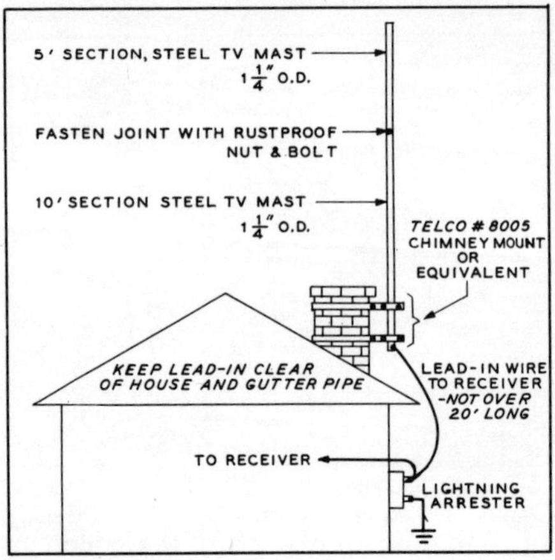

Fig. 5 If you are surrounded by obstacles, you might consider a vertical antenna mounted to the chimney of the house. Simple TV antenna clamp may be used to hold antenna in place. Use insulated wire for lead-in, and keep it well clear of the gutter pipes.

It is simple to erect and works well over a limited frequency range, such as a single amateur band, or a shortwave broadcast band. It has a non-directional pattern, and is responsive to atmospheric and man-made noise in the same manner as is the simple vertical antenna. Dimensions of "ground planes" for various bands are shown in Figure 6. Four horizontal *radial wires* are placed beneath the antenna section. These wires act as a resonant ground screen for the vertical antenna and greatly improve reception, particularly at the higher frequencies. The radials run away from the base of the antenna, though it is not necessary that they travel in straight lines. The usual procedure is to weave them around the house, through fences, and through the garden to make them as unobtrusive as possible. The radials may slope down from the base of the antenna, but they should not project above the base level of the vertical section. The ground plane is relatively unconcerned about its location. It may be strapped to a chimney or a short pole, or placed atop a garage. As with any other antenna, it should be removed as far as possible from power, telephone, and utility wires.

This installation employs a transmission line to conduct the signal from the vertical antenna to the receiver. A length of 52-ohm coaxial line should be used. The outer shield of the line is attached to the four radials, and the center conductor is attached to the vertical section. The use of the coaxial line prevents signal pickup by any part of the antenna other than the vertical section, which is a great advantage over the simple vertical antenna of Figure 4. Noise pickup is greatly reduced, and high frequency operation is improved.

The "Fan" Vertical Antenna

It is possible to obtain frequency coverage over the entire 3.0 to 30.0 mc spectrum with the *Fan Vertical antenna* illustrated in Figure 7. Half-dipole antennas are arranged in the form of a fan and are connected in parallel at the lower end. At resonance, each dipole section makes an approximate

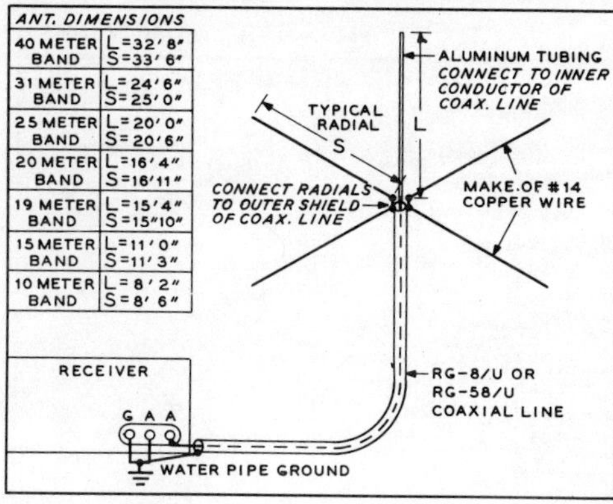

Fig. 6 The ground plane antenna is a good performer over a small frequency band. It must be cut to the listening range for optimum results. Shown at right are antenna and radial dimensions for the most popular shortwave bands.

match to a low impedance coaxial line. The sections that are off-resonance have small effect upon the operation of the tuned section.

The fan antenna may be connected to the receiver by a length of 50 ohm coaxial line. This line may be buried underground, if desired. The outer shield of the line is grounded at the antenna end of the system by a short length of pipe driven into the ground at the point the coaxial line emerges. The shield of the line is attached to this pipe. The junction of the fan wires are attached to the inner conductor of the coaxial line. This antenna is relatively non-directional, but is more responsive to man-made noise and ignition interference than is a horizontal antenna of the same type to be described later in this chapter.

The Dipole Antenna

The *dipole antenna* is a center fed, tuned antenna, intended for optimum receiving results over a small band of frequencies. Doublet antennas are used for reception (and transmission) by many amateurs, and are also

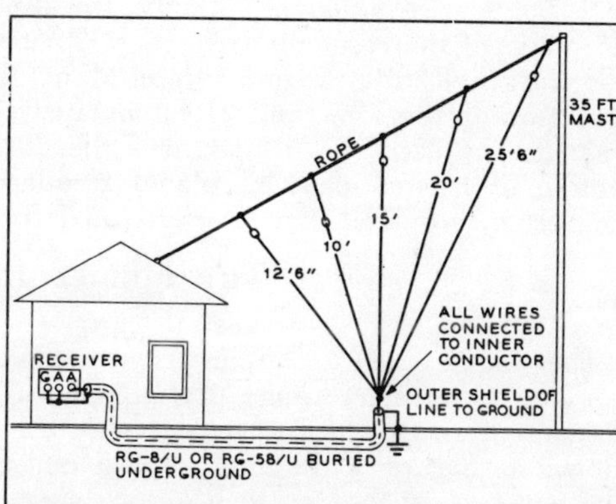

Fig. 7 "Fan Vertical" antenna is multiple system providing good reception over a wide range of frequencies. Place the antenna clear of buildings for best results.

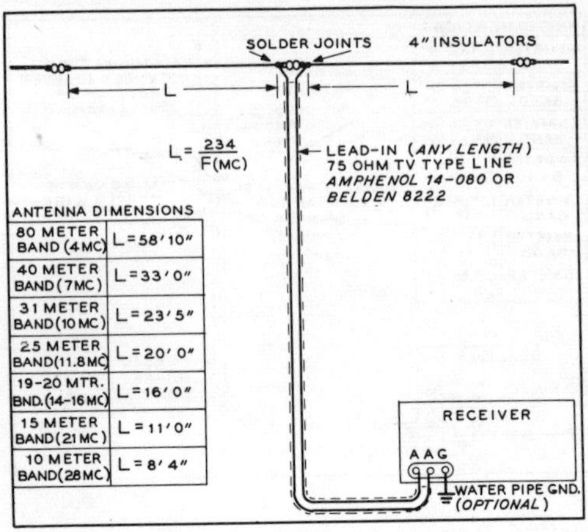

Fig. 8 The dipole antenna is "old standby" for good results over narrow shortwave region. Several dipoles may be erected to provide coverage of most important shortwave bands.

used for commercial shortwave circuits. The length of the flat top portion of the doublet determines the region of the shortwave spectrum to which it is tuned. The doublet is split in two at the center by a small insulator whose size is not critical. Each wire of the flat top is attached to one conductor of a two wire transmission line. Special balanced 75 ohm polyethylene insulated transmission line intended for use with dipole antennas is made by several companies and can be purchased in any large radio store. The length of the transmission line is not critical. It should be long enough so that it can drop away from the antenna at right angles, as shown in Figure 8.

The wires of the transmission line are attached to the two antenna terminals of the receiver. No ground connection need be used. If the receiver has only one antenna terminal, one wire of the transmission line should be attached to it, and the other wire fastened to the ground terminal of the receiver.

The dipole antenna tunes rather sharply, and optimum results are obtained only over the frequency range for which it was designed. Reception outside the operating range of the antenna will be degraded, as compared to a single wire antenna. Luckily, the dipole will work well at the third harmonic of the resonant frequency. That is, a dipole tuned to 7 mc, will also work well in the 21 mc region. Many amateurs use such an antenna for operation on the 7 mc and 21 mc amateur bands. A dipole tuned to 10 mc may also be used for reception of the 28-29.7 mc amateur band. For reception of harmonically unrelated frequencies, two or more dipoles may be erected, provided they are separated from each other by a few feet.

The Folded Dipole

The *folded dipole* is an improved version of the simple dipole. It will work over a greater range of frequencies, and will effect a more efficient transfer of energy to the average shortwave receiver. The folded dipole is shown in Figure 9. It employs two end-connected wires in the flat top. The wires are spaced an inch or two apart, and one of the wires is broken

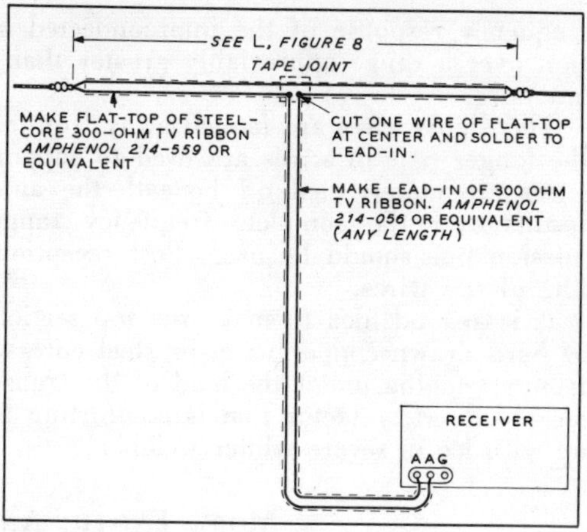

Fig. 9 Folded dipole made of 300-ohm "ribbon" line has a greater frequency span than the simpler antenna of Figure 8. Length L is twice the dimension given in Figure 8.

at the center and attached to the transmission line.

An inexpensive version of the folded dipole may be constructed from 300 ohm TV-"ribbon line" as shown in the illustration. For greatest strength, the horizontal portion of the antenna should be made of "ribbon" having steel-core copper wire, otherwise the antenna is liable to break at the center joint in a high wind. The wires of the transmission line are attached to the receiver terminals as in the case of the ordinary doublet antenna.

The Triple Dipole

It is possible to combine three dipoles into one antenna to form an "all-wave" antenna system covering the range of 5 mc to 30 mc. Good reception may even be had over a wider range of frequencies if a small loss in signal strength is accepted. A typical *triple dipole* antenna is shown in Figure 10. The flat top is composed of three separate dipoles connected to each other and to the transmission line at the center of the antenna. The combined

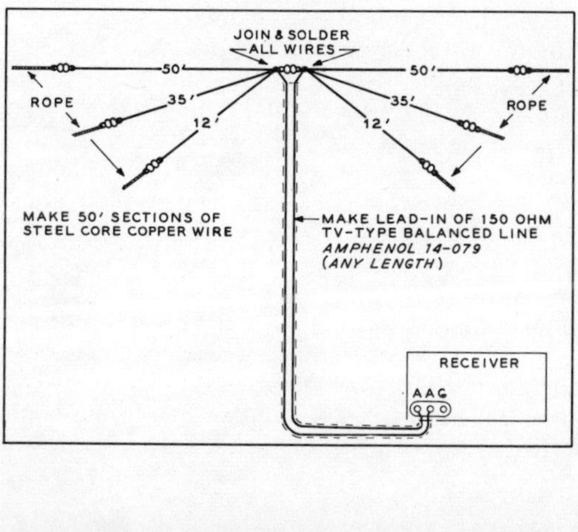

Fig. 10 Triple dipole covers broad frequency range and is recommended for general listening. Erect antenna as high in air as possible for optimum results.

frequency response of the interconnected antennas results in good reception over a range appreciably greater than that provided by two separate antennas.

The dipole wires are fanned away from each other. In most installations, the longer pair of wires are used to support the antenna, and the shorter wires are merely tied off beneath the antenna, as shown. For optimum results over the complete frequency range, a 150-ohm balanced transmission line should be used. Best reception is had at right angles to the line of the wires.

It is a good idea to make the top sections of the antenna from lengths of hard drawn copper wire, or steel core wire. This will prevent the wire from stretching under the load of the transmission line, and also will give an added safety factor that is comforting to have when the antenna coats up with ice in severe winter weather!

More Exotic Antennas

The ultimate reception is to be had with the use of beam antennas. These antennas provide optimum results in a certain direction, and must be "aimed" at the transmitting station. The usual beam antenna has a pickup lobe varying from twenty to seventy degrees wide permitting the antenna to cover a large area of the world. For example, a beam antenna user in New York could cover the whole of Europe with one high gain antenna. These specialized antennas minimize the fading and "flutter" of a long distance signal, and reduce the interference received from a signal that is "off the beam." The *Beam Antenna Handbook* and the *VHF Handbook* cover antennas designed for amateur transmission and reception on the high frequency and very high frequency bands. The reader is referred to these texts for the "real dope" concerning these interesting antennas. Both books are published by Radio Publications, Wilton, Conn., and are available at your local radio distributor.

CHAPTER VIII

Reception Techniques

Shortwave reception is an art. There's more to it than rushing out and buying a receiver, hanging a piece of wire on it for an antenna, and immediately hearing exotic DX stations and having your mailbox flooded wtih verification cards! Sure, you can hear the VOA, or the BBC with a piece of damp string tied to your receiver for an antenna. But if you are looking for real DX—an amateur on the opposite side of the globe, or a weak broadcast station in Outer Slobbovia—it is necessary to absorb a few tricks of the trade. Here are some pointers that will start you on the exciting road to better DX reception.

You've got to know *how* to listen, *when* to listen, and *where* to listen! You also should have a good working knowledge of radio propagation and its effects on DX signals. Armed with these facts, you have a pretty good chance of hearing the station you are looking for. Haphazard scooting about the dial will provide a lot of fun and net some interesting stations, but the serious listener tracks his quarry like a bloodhound. Once he finds his prey he goes after the all-important QSL card with equal vigor. You, too, should "make like the bloodhound" for maximum enjoyment of this great hobby.

Listening Hints

Since the ardent SWL and would-be amateur spends a good many hours at the receiver, he should try to make himself as comfortable as possible. He should make the job of receiver tuning simple, with a minimum of fatigue. The receiver can be placed in the middle of a large desk or table, slightly above elbow height when the operator is seated. The receiver is pushed back from the front edge of the table sixteen inches or so in order to allow the listener to rest his elbows on the table. The operator should sit facing the receiver, with illumination coming over one shoulder. A lamp on

Rare and exotic QSL cards from all over the world may be yours in return for sending a reception report to the many shortwave broadcasting stations. Be sure to include time of reception, wave band, selections heard, and your name and return address. Your cards will make a wonderful collection to be proud of!

top of the set gives a poor light, as it usually hits the eyes and leaves the tuning dials of the receiver in a shadow. Less light will be required if the the dials of the receiver are illuminated.

The serious DX'er uses earphones rather than a loudspeaker. Weak and semi-intelligible signals can be understood more readly when earphones are used in preference to a speaker. Room reverberation and external noises limit shortwave speaker reception to loud and steady signals. Good quality, light weight earphones of 2,000 ohms impedance will work with the majority of shortwave receivers. War surplus 'phones having sponge rubber earpieces are also satisfactory. Make sure the headband fits your head, and that the earpieces feel comfortable. Many are the hams and SWL's who walk about with deep grooves worn in their noggins, from untold hours of listening with ill-fitting "light-weight" earphones!

Operating Aids

All-important operating aids are a notebook, a pencil, and a clock. The notebook is a "must" for logging program information, dial settings, and frequency notes. The clock should be an electric one, set with radio time signals from WWV or some other accurate source. You'll need exact time so your verification data can be correct. The pencil is handy for writing in the log, and for chewing when your best DX reception is blotted out by QRM.

The log book should be used to keep a record of important facts. You should log the call letters or identification signal of the station, the time of reception (local time and GMT), program material, the signal strength, fading of the signal, and interference (if any). Other pertinent facts may be jotted down in the book at the moment you note them. The important thing to remember is that you should keep a *permanent* record of reception, as you might need it to refer back to it in the future. If you have written your notes on the back of an old racing form, they probably will end up in the wastebasket in a few days. Notebooks are cheap, and it's better to be safe than sorry! Buy an inexpensive book, and write down all your observations. You'll be glad you did, as you will create a record of your achievements that will be interesting to look over in later years.

Another important operating aid is an accurate list of all the shortwave broadcast stations, listing call letters, frequencies, and hours of operation. The *World Radio Handbook,* available through your local radio store is recommended, since it provides extensive coverage of all the important shortwave broadcast stations throughout the world.

Finally, it would be a smart move to join one of the many good shortwave listener's clubs located in the United States and in many other countries about the world. For additional information about club activities and bulletins you might contact the *Newark News Radio Club,* 215 Market Street, Newark 1, N. J. To keep up with the changing world of shortwave radio several magazines run SWL columns, giving DX-tips and tuning information. One of the best of these columns may be found in *Popular Electronics* magazine each month.

How, When and Where to Listen

Almost every piece of shortwave equipment you might wish to use can be purchased if you have enough money. You can read books and learn all you wish to know about antennas, receivers, and reception. You can hire a fellow to put up your antenna and align your receiver for you. *But the final results you obtain are up to you!* The technique of DX reception can be learned only by experience, coupled with patience. Spare time spent in listening and hunting for DX, and in obtaining the "feel" of the band will turn the greenest tyro into an experienced DX-chaser!

You must learn that not all weak, fluttery signals are DX. Learn to distinguish between the weak foreign signal and the watery "backwave" received from a skip-zone signal a few hundred miles away. Acquire an idea of the signal strength of some of the more powerful overseas stations so that you may use them as "light houses" to judge propagation conditions over a certain path. Just because *Radio Australia* is below par in signal strength does not mean that the less-powerful *Radio New Guinea* a few hundred kilocycles away will not be heard. The ionosphere is a mysterious device, and the strength of one signal usually bears no meaning to the strength of other signals. However, if the majority of signals from a certain direction are weaker than usual, the chances are that poor propagation conditions over this particular route will affect all signals. When the "old standbys" are putting in powerful signals it is an indication that conditions are good, and yet relatively stable. Freak reception will be at a minimum, but on the other hand you will stand a better chance of finding the weaker stations in the general period of good propagation.

You must learn what areas of the world come through at certain times of the day and night on particular frequencies. Listen to a powerful station such as the BBC that transmits the same program material on a number of frequencies. Switch your receiver back and forth between the signals. You can get a good idea of which frequency band is best for reception from the area in which the station is located. Be sure you know the source of the signal you are examining! *Radio Moscow,* for example is rebroadcast by satellite countries, and also by repeater stations in remote Siberia, and no identification of transmitter location is given. The language used in the broadcast provides no clue either, as most countries transmit many hours of foreign language programs. Careful listening, however, will log many stations that are consistent in frequency and programming enabling them to be used as marker stations for future reference. Mark them down, along with their frequencies and time of reception, so that you can look for them in the same spot another day.

Auroral "Flutter" Signals

Radio signals reaching you via great circle paths that cross the polar region suffer from auroral effects, and generally have a rapid "flutter," or "echo" on them. A listener on the East Coast of the United States, therefore, searching for stations in the Orient or the Far East should pay attention to signals that exhibit this effect. Western DX'ers, on the other hand, would observe this unusual flutter on European signals passing near the North Pole.

Signals passing through the South Pole auroral area are subject to the same type of distortion. Southern Argentina, the Falkland Islands, and the Antarctic stations can be quickly spotted by the peculiar fluttering of the carrier wave. Learn to recognize the auroral flutter signal—it means DX from across the polar region of the world!

"Long Path" Signals

There is nothing to prevent radio signals from reaching you via the long transmission path. That is, one directly opposite to the great circle route. Such propagation does occur, principally in the 6 mc to 25 mc range. Late afternoon signals from Australia and the Far East will often arrive on the Eastern Coast of the United States passing across Africa. European signals can be heard with surprising strength on the West Coast during the morning hours as they come across Australia and the Indian Ocean. These signals avoid the auroral zone, and the characteristic flutter is absent. In addition, signals from India and Central Asia may arrive via the southern path across the Antarctic continent, or across the North Pole. On many occasions, the signals may arrive by both paths. The time lag between the two different propagation paths results in a bad "echo" which usually reduces the intelligibility of such a double path signal to zero. Long path signals are often extremely strong, especially during the fall and winter months.

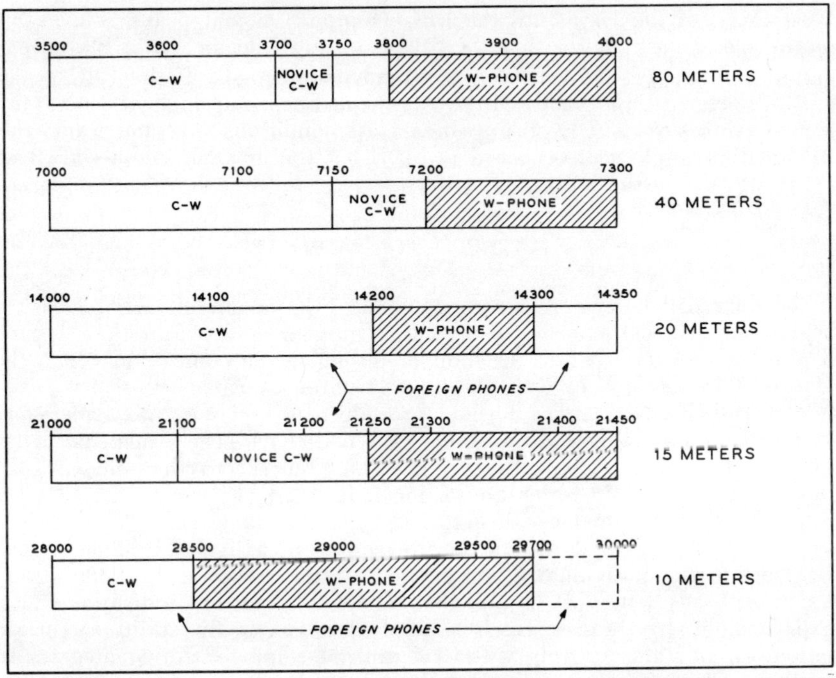

Radio hams occupy small segments of the shortwave spectrum, as shown in this chart. Listen near the edges of the American phone sections for DX stations!

Hunting DX in the Amateur Bands

Many countries that boast no broadcast facilities can be found in the amateur bands. In addition, expeditions (Operation *Deepfreeze,* for example) can usually be heard operating within the amateur bands. A tabulation of the amateur frequencies is given in Figure 1, with the American 'phone portions of the bands noted. Phone DX stations are usually found *outside* the limits of the American 'phone band, since they find they cannot compete with the high interference level found in the W-portion of the band. Stations outside the Continental United States which are under the jurisdiction of the Federal Communications Commission (Guam, Hawaii, Puerto Rico, and other islands) operate within the limits of the American phone band.

A clue to the propagation conditions may be obtained by monitoring the local amateur phones operating in the 10, 15, and 20 meter bands. When these stations are working and calling DX, it is a sure sign that the band is "open." Since the amateur stations do not run the high power levels permitted the shortwave broadcasting stations, and since their antennas are smaller, the ham signals do not equal the strength of the b-c stations. In addition, the lower powered signals must "ride with the skip," taking full advantage of the optimum propagation conditions. For instance, the 21 mc (15 meter) band does not open for amateur work between Europe and the

West Coast of the U.S. until the BBC (London) signal peaks at over S9 on the West Coast. By the time the BBC is up to 20 decibels over S9, propagation will support the relatively weak amateur signals. If the BBC peaks to 40 db over S9, the band will be alive with European amateurs, many of them having surprisingly strong signals. As conditions wax and wane, the BBC station might still be heard at S8-9, but the amateur signals are lost by absorption in the ionosphere.

Amateur Call Letters

Amateur call letters are assigned by the FCC on an impartial basis. A typical amateur call consists of a letter, a number, and two or three letters. The letter or letters before the number designate the country in which the station is located (W or K in the case of stations located within the continental United States). The normal procedure is to issue "W" call letters in a certain area. When all available "W" calls have been issued, the letter "K" is then employed as the prefix. The numeral in the amateur call designates the area in which the station is licensed (Figure 1). The United States is divided into ten call areas, bearing the numerals 1 to 0 (zero). Following the numeral are two or three letters assigned in alphabetical sequence which individualize the amateur station. These suffix letters may begin with any letter of the alphabet except "X," an old indication of an experimental-type license which is no longer issued. In addition, certain letter combinations starting with "Q" are not employed for amateur calls, as these combinations are used in the international *Q-signal* code groups.

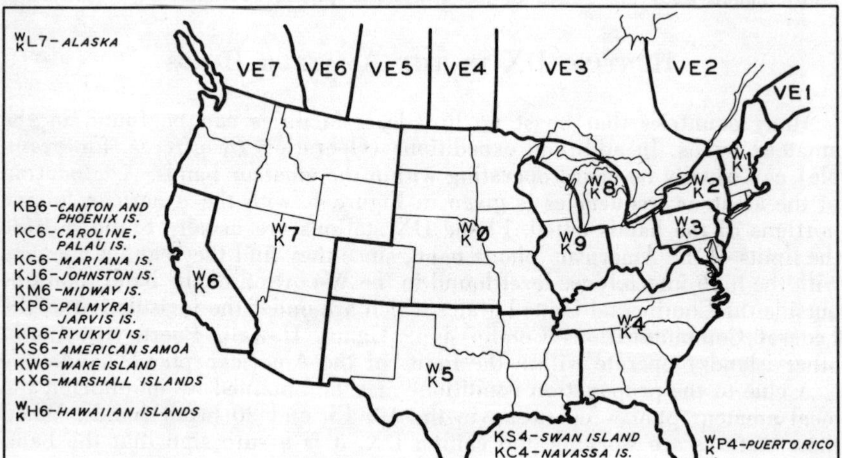

Fig. 1 United States is divided into call areas. Amateurs within these areas may be identified by the numeral in their call. The Radio Amateur Callbook lists the addresses of U.S. and overseas amateurs, and is a great help in writing for amateur QSL cards. Search the ham bands for rare DX!

Foreign call letters begin with a letter or a group of letters that usually identify the country, followed by a number and one, two, or three suffix letters. In some instances, a whole group of islands may be identified by the same prefix, so it is not always a simple matter to spot the location of the station merely from the prefix. For example, the prefix *VP2* covers both the *Leeward Island* group and the *Windward Island* group in the West Indies. An amateur station with this prefix might be located on any one of a dozen or more islands. Another famous example is the prefix *VP8*, covering the Falkland Islands, the South Orkney Islands, the South Shetland Islands, South Georgia Island, and the British Antarctic Zone. The moral of this is not to jump to conclusions when you hear an amateur call. Look the prefix up on a country list. You *may* be fooled.

Amateur Operation

An interesting group of stations for the DX-hound to hunt are the *maritime mobile* stations, operating in the 10 and 15 meter amateur phone bands. These sea-going hams are usually members of the crew of a freighter or oil tanker, and they visit the remote corners of the world. Federal law requires

The popular programs of the British Broadcasting Corporation may be heard throughout the world in all of the shortwave broadcast bands. Shown above is one of the powerful transmitters located in the British Isles that has made the beautiful tones of "Big Ben" well-known to all shortwave listeners. All major countries are represented by powerful shortwave stations such as this one.

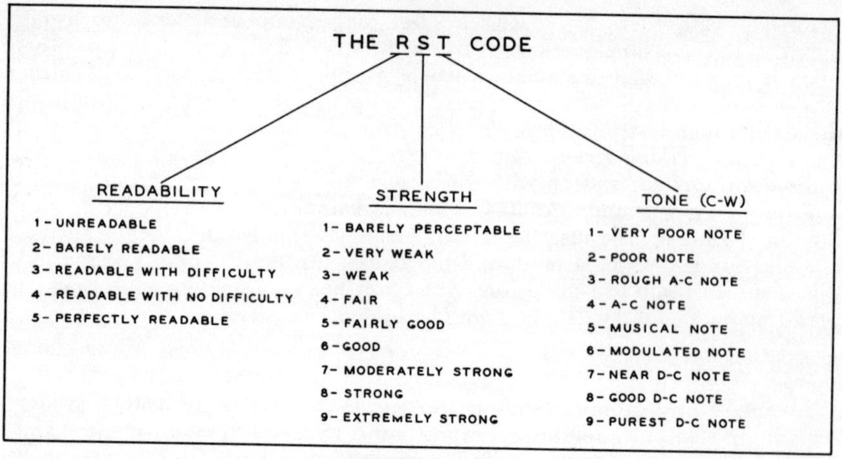

Fig. 2 The RST code is used by amateurs to permit rapid and concise transmission of signal reports. Use this form of report when QSL'ing to hams.

that the exact location of the ship be given, and you might be lucky enough one day to hear Captain Kurt Carlsen of the *Flying Enterprise II* announce, "*This is W2ZXM, maritime mobile, at 118 degrees west longitude, 35 degrees north latitude, in the China Sea.*" A peek at your atlas will pin-point the location of the vessel.

Some amateur stations are engaged in handling messages between Armed Forces personnel overseas and their relatives at home. Daily contacts are maintained by many hams with the Antarctic bases scattered across the south polar continent and with other bases in Europe and Africa. Other amateurs are DX-chasers like yourself, and specialize in tracking down and contacting rare and unusual stations. You can locate GI hams in dozens of countries by monitoring these message handling stations.

Amateurs have their own language, which is a distant relative of the Morse telegraphic code abbreviations. These old abbreviations are often used "as-is" for phone work, along with modern ones thrown in for good measure. A little listening experience will enable you to "translate" this jargon into good English.

The SWL should remember one important fact when he tries to obtain a QSL card from an amateur station he has heard. The amateurs do not *ask* for SWL reports. Most shortwave broadcast station solicit reports and are morally obligated to reply to them. The ham seldom does this, and therefore the SWL takes a justifiable risk when he sends his card to an amateur station. Many hams do not reply to SWL cards, since they cannot afford the postage bill. Be sure to include an International Reply Coupon with each SWL report to an amateur if you expect to receive a reply.

The C-W Amateur

Unfortunately, most shortwave listeners are eager to log phone amateurs, but few seem to pay much attention to DX hams that operate c-w. The author has been an active amateur for over twenty years and has received literally thousands of SWL cards and letters reporting reception of his phone signal, but only a handful of letters confirming his c-w transmissions. None of the c-w reports came from the United States, either! It seems like the American SWL's shun the code more than their overseas brothers! In other parts of the world the reverse is true, since a few European countries require the SWL to obtain a license to *listen.* If he is further interested in obtaining an amateur license to transmit, he must present a certain number of QSL cards received from c-w amateurs whose signals he has heard. Consequently the c-w ham receives SWL cards from foreign listeners, but *almost never* from SWL's in the U.S.A. This is a pity, since the chances of obtaining a reply from a c-w report are many times greater than from a phone report, because a SWL card is much more of a novelty to a c-w station. Many phone amateurs are swamped with SWL cards, and ignore them entirely. On the other hand, low powered foreign c-w hams are anxious to obtain reports on their signals, and a SWL card means much to them.

If you are interested in someday obtaining your ham "ticket" you must master the code. Why not "kill two birds with one stone" by copying c-w ham stations? You can find real DX, pick up interesting and rare QSL cards and improve your code copying ability all at the same time! Over 75 percent of the amateurs outside of the United States operate c-w, so you will find unusual DX and rare countries in the c-w portions of the amateur bands.

Your SWL Report

Hearing the station is only half the battle. The other half is obtaining a confirmation of your reception. Some stations are prompt in replying, and a few never reply at all. One or two require that you send them postage. Some reply to a few reports and ignore others. By and large, the majority of broadcasters try wholeheartedly to verify all the *accurate* reports of reception that they receive. Here are a few "tricks of the trade" that will help you increase the percentage of your replies.

What should your report consist of? You might writer a letter to the station, or you can send a shortwave listener card *(SWL card)*. The card may be a home-made one, or it may be printed by a QSL card printer. Examples of some printed cards are shown in the illustration. Whatever form of report you employ, it must include some rather pertinent information if you expect the report to be of any use to the station. First of all, you must give your *name* and *address.* Don't laugh! A lot of fellows forget this all-important bit of information. Put your *full* mailing address on the card or letter as well as on the envelope just in case the envelope is put aside. Secondly, you should state the *date* and *time* of reception. Be sure to specify *what kind of time* you are using (Eastern Standard Time, Daylight Savings Time, GMT, etc.). Name the *operating frequency* of the station, or at least the approximate waveband of reception. If your report is sent to a broadcast station list the *names of the musical selections* (if you know them), or at least the *type of music* being played. Note any *announcements* you hear, and any other program information that you can obtain. Next, you should give a *reception report,* indicating the *signal strength* of the station. This report should be given in the RST code, shown in Figure 2. A phone signal may perhaps have a report of R4 S7, indicating a moderately strong signal, readable with practically no difficulty. Receivers having signal strength meters calibrated in S-units can give a close approximation of the received signal in relation to other signals on the band. The "tone" section of the RST code is used for c-w signals. It is the custom to group the report for such signals, as RST 558, for example. This indicates a perfectly readable signal of fair strength, having a pure note.

Finally, you should indicate the *type of receiver* you are using (number of tubes, etc.) and the *type of antenna* (length, height). A complete report will also indicate the weather condition and the general state of radio reception at the time the report is made.

There is an old saying: "You can catch more flies with honey than you can with vinegar"—remember this. In return for your information, you are asking the station to spend its money to have verification cards printed and mailed. The printing and postage bills for a popular station can run into many hundreds of dollars a year. A bunch of worthless reports will make the station owner think twice before he shells out his money for verification cards! Make your report as painstaking and as neat as you can. In order to relieve the financial drain upon the station, you should enclose an *International Reply Cupon* with your report. This cupon is available at all post offices for a few cents, and may be converted into sufficient postage for one piece of first class mail in most foreign countries. The IRC cupon will boost your percentage of replies, so use it!

The length of time required for a reply to a report varies widely, depending upon the policy of the station, and the speed of the mail. The average time to process a report and mail a verification card is about three months. Many stations reply in less time, but a few mail out listener reports only once a year! After an interval of six months or so, it might be a good idea to mail a second report to the station if you have received no reply, since your original report or the reply to it may have become lost en route. On the average, a return of about 70% may be expected with no particular strain or pain on your part.

Amateur QSL Cards

Obtaining QSL cards from foreign amateur stations presents other interesting problems. The report to the ham should be as all-inclusive as your report to the s-w broadcasting station. You may obtain the address (QTH) of most of the foreign hams from the latest edition of the *Radio Amateur's Callbook*, available at most radio supply houses. This will enable you to send your report of reception directly to the amateur. Be sure to incude an IRC coupon with your report. Certain countries—particularly the "Iron Curtain" ones—are a little "touchy" about their amateurs receiving direct mail from overseas. The hams in these countries have therefore set up QSL clearing houses which operate from a postal box, usually located in the capital city of the country. All incoming and outgoing QSL cards must pass through this semi-official clearing station. A SWL card or letter sent to "Amateur Radio Station so-and-so" in the Soviet Union would either end up in the clearing station, or in the wastebasket of the censor.

Unfortunately, the hams behind the "Iron Curtain" cannot or will not reply directly to QSL and SWL cards. All replies pass through their QSL

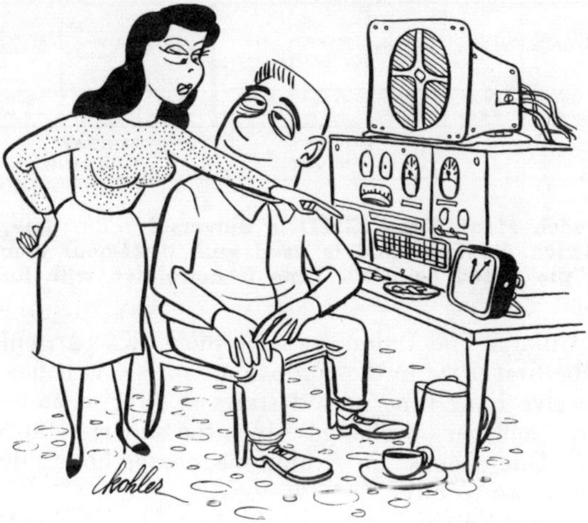

"So it's only 8 p.m. in the Fiji Islands! Get out of your dugout canoe and come to bed!"

bureau to the QSL bureaus of other countries. The amateur QSL bureaus of W-land do not handle SWL cards, and unless special arrangements can be made through one of the larger SWL clubs, your chances of receiving a verification card from a ham in one of these countries is pretty slim.

Save Your Cards and Reports!

As you become an experienced listener, your stack of verification cards will grow. Be sure to save them! A large photographic album is a fine place to exhibit the cards, and offer them protection at the same time. Two prized cards in the author's collection of over 4000 QSL's are the first SWL verification cards he received in 1930.

Radio Time

Since Columbus proved he could reach the east by sailing west mankind has run into a number of problems arising from the fact that the earth is round. One of these problems intimately concerning the SWL and the amateur is the fact that the local time is not the same at different points around the earth. When it is midnight in Los Angeles, it is 3 a.m. in New York, 8 a.m. in London, 11 a.m. in Erevan, noon in Teheran, 4 p.m. in Manila, 7 p.m. in Noumea, and 10 p.m. in Honolulu. In addition, it might be Sunday morning in New York, while it is Saturday evening in Manila!

As you can imagine, computing the time in a particular part of the world could turn out to be a bit tricky, especially since some countries as Ceylon,

GMT TIME CHART											
CALCUTTA	SINGA-PORE	MANILA	TOKYO	SYDNEY	WAKE IS.	FIJI IS.	HAWAII	FAIR-BANKS	SITKA	PACIFIC STANDARD TIME	MOUNTAIN STANDARD TIME
0600 +6 H	0700 +7 H	0800 +8 H	0900 +9 H	1000 +10 H	1100 +11 H	1200 +12 H	1300 −11 H	1400 −10 H	1500 −9 H	1600 −8 H	1700 −7 H
MONDAY					MONDAY NIGHT ←	DATE LINE		→ SUNDAY MORNING			SUNDAY

Fig. 3 Greenwich Mean Time (GMT) is universal radio time. It is based on time in Greenwich, England and is used with a 24-hour clock. No notations of "am" and "pm" need be used. Time is correlated with longitude degrees.

India, British Guiana, and Indonesia have their own particular time. Saudi-Arabia takes the first prize in the confusion contest, as it has no local standard time. A twelve hour time period starts at sunrise and sunset, and the starting time of each period changes since the sun rises at a different time each morning! "One o'clock" in Arabia means one hour after local sunrise, and differs from day to day.

Some years ago the British helped to solve this knotty problem by adapting a 24 hour clock, and establishing time zones throughout the world based upon this clock. In addition, the new time system had nothing to do with the local time, as it was based upon the time of a standard clock located in Greenwich, England. This system is called *Greenwich Mean Time* (GMT) and is not so nutty as it sounds. After all, our conception of time is purely arbitrary. We could get along just as well if our days were divided into thirteen hours instead of twenty-four. We could make each day fifty hours long by making each hour shorter. The concept of "a.m." and "p.m." is also artificial, since the clock hands point only to numbers. We mentally add the idea of "afternoon" or "morning" to the time we read on the clock. Greenwich Mean Time does away with all this balderdash in one drastic revision. Since Greenwich is located on the zero degree meridian, GMT is tied directly to degrees of longitude. Thus, if you know where you are, you know your time compared to the time in Greenwich, England. The position of the sun in the sky has nothing to do with your local Greenwich time. If your location happens to be *east* of Greenwich, your local time will be *ahead* of GMT, and if you are *west* of Greenwich your time is behind GMT. In either case, you can express your time in Greenwich hours.

The 24-Hour Clock

GMT is based upon a 24 hour clock, instead of the usual 12 hour clock we are familiar with. Thus, in the GMT system 9 a.m. is 0900 hours, 1 p.m. is 1300 hours, 5:30 p.m. is 1730 hours, and 11:55 p.m. is 2355 hours. The notations of "a.m.", "p.m.", and "minutes" are not required. Since GMT is based upon the idea of longitude, any point in the world may reckon its time relative to GMT *if* the degree of longitude is known. New York, for instance is west of Greenwich and its local time is 5 hours behind GMT. 8 a.m. in New York is therefore 1300 GMT. Singapore is east of Greenwich, and its local time is 7 hours ahead of GMT. 4 p.m. in Singapore is thus 0900 GMT. This simple (?) situation is summarized in Figure 3. The important thing to remember is that if you express your local time of

GMT TIME CHART											
CENTRAL STANDARD TIME	EASTERN STANDARD TIME	ATLANTIC STANDARD TIME	RIO DE JANIERO	AZORES IS.	DAKAR	(GMT) LONDON	CENTRAL EUROPEAN TIME	(MSK) MOSCOW	ADEN	SVERD-LOVSK	BOMBAY
1800 *−6H*	1900 *−5H*	2000 *−4H*	2100 *−3H*	2200 *−2H*	2300 *−1H*	2400 *0*	0100 *+1H*	0200 *+2H*	0300 *+3H*	0400 *+4H*	0500 *+5H*
SUNDAY					SUNDAY NIGHT		MONDAY MORNING				MONDAY

If you know where you are, you know your time compared to GMT. If you are east of Greenwich, your local time will be ahead of GMT, and if you live west of Greenwich, your time is behind GMT, as shown in the above table.

reception in GMT, the station to whom you send a report will be immediately able to translate the time of reception into its own local time. It can then cross-check your report easily. In fact, many shortwave stations and amateurs use GMT as their "station time" instead of their own local time.

Figure 3 shows a conversion chart for all the time zones of the world. Look up your time zone and convert your time into GMT. You can also tell GMT time in other areas of the world from this chart. Learn to use it. It will be a big help in aiding you to translate world time systems.

Time Signals

Many stations throughout the world transmit time signals on various frequencies. The stations you will probably use the most are the *National Bureau of Standards Stations WWV* and *WWVH* located in Maryland and Hawaii respectively, and the *Canadian Dominion Observatory Station CHU* in Ottowa. *WWV* and *WWVH* announce the time every five minutes, beginning on the hour, and *CHU* announces the time every minute. Listeners in the Pacific area can use *JJY*, Tokyo or *WWVH*. A table of time signal broadcasts is given in Figure 4.

In addition, stations *WWV* and *WWVH* conduct a radio propagation warning service. Each twenty and fifty minutes past the hour these stations transmit a code signal indicating the propagation conditions on the North Atlantic and North Pacific paths. These signals consist of the letter N, W, or U followed by a number as explained in the drawing. Check *WWV* or *WWVH* for a quick summary of radio propagation conditions!

The Shortwave Broadcast Bands

Jumping from broadcast reception to shortwave reception requires quite a reorientation on the part of the listener. No longer are the stations spaced at modest 10 kilocycle intervals across the band. Instead, the shortwave stations are packed into the spectrum in what seems to be a haphazard fashion. The listener will find broadcast stations intermingled with code transmissions and with each other in an apparent state of chaos. However, there is some method in the apparent madness, and most shortwave broadcast stations are grouped into broadcasting "bands," or segments of the spectrum. These areas are relatively free of code and radio teletype interference, and here you will find the majority of DX broadcast stations. The bands are the 90 meter band (3200-3400 kc), the 60 meter band (4750-5000 kc), the 49 meter band (5900-6,400 kc), the 41 meter band (7,100-7,400 kc), the 31 meter band (9,200-9,700 kc), the 25 meter band (11,600-12,000 kc), the 16 meter band (17,700-17,900 kc), the 13 meter band (21,450-21,750 kc), and the 11 meter band (25,400-26,100 kc).

A quantity of broadcast stations operate outside the limits of these bands, however you will find that the majority of shortwave broadcasters stick pretty close to these internationally recognized regions.

Each shortwave broadcast band has its own peculiarities of propagation, depending upon the time of day, the season of the year, and the sunspot cycle. Let's look at these bands and see what general conclusions we can reach about them.

The 90 and 60 Meter Bands

The 90 and 60 meter bands are primarily used for local coverage broadcast work in the tropical areas. During the evening hours of the winter months it is possible to hear many South and Central American stations in these channels. The high static level reduces the utility of the 90 meter band during the summer months, but the 60 meter band is useful the year around. The "lighthouse" of the 90 meter band is *ZQI* in Kingston, Jamaica, operating on 3360 kc during the early evening hours.

The 90 meter band is almost exclusively populated by South and Central American stations, but the "DX-bloodhound" will find many 60 meter African signals from the Congo, Angola, Kenya and the Union of South Africa wedged between the Latin signals. In the early hours of the morning you might hear some of the Far East broadcast stations, such as Jogjakarta, Indonesia, and Columbo, Ceylon. *Radio Moscow* is also very active in this region, as is *All India Radio* in Calcutta.

The 49, 40, and 31 Meter Bands

These three bands form the "backbone" of shortwave broadcasting, as they contain a majority of the s-w stations of the world. Over a hundred countries compete for your ears amid this beehive of electronic activity. These bands are all-year performers, and are "open" most of the hours of the day and night, with conditions peaking just before sunrise, and during the evening hours. The static level is much weaker than on the lower frequencies, and the main interference comes from the multitude of signals. As many as eight or ten stations operate on the same channel during the hours of the day or night. If propagation is right, you may hear several stations from different parts of the world at the same time on the same spot on the dial. The din that is created is amazing!

The "lighthouses" on these bands are the *Voice of America,* the BBC *Radio Moscow,* the *Canadian Broadcasting Commission, Radio Australia,* the *Voice of Brazil,* and others too numerous to mention. Buried beneath the "big" signals are a wealth of weaker stations. The experienced DX-er finds this his "happy hunting ground," and he combs the bands avidly for signals from *Radio Monte Carlo,* Karachi, Pakistan, Brazzaville, the Gilbert Islands, Ceylon, Tahiti, Austria, Greece, Syria, and a host of other exotic places. Often, the radio conditions conspire to project a station above its comrades on the dial. A fluctuation of the ionosphere and *Radio Sweden*

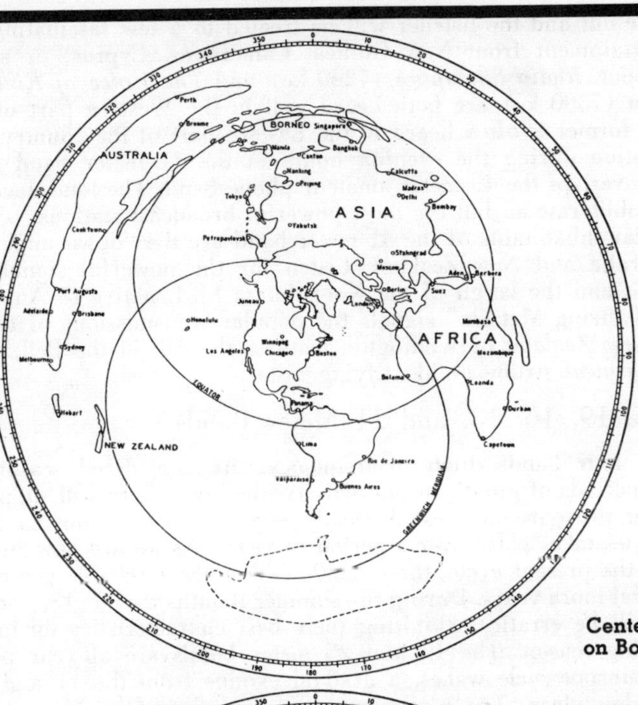

Centered on Boston

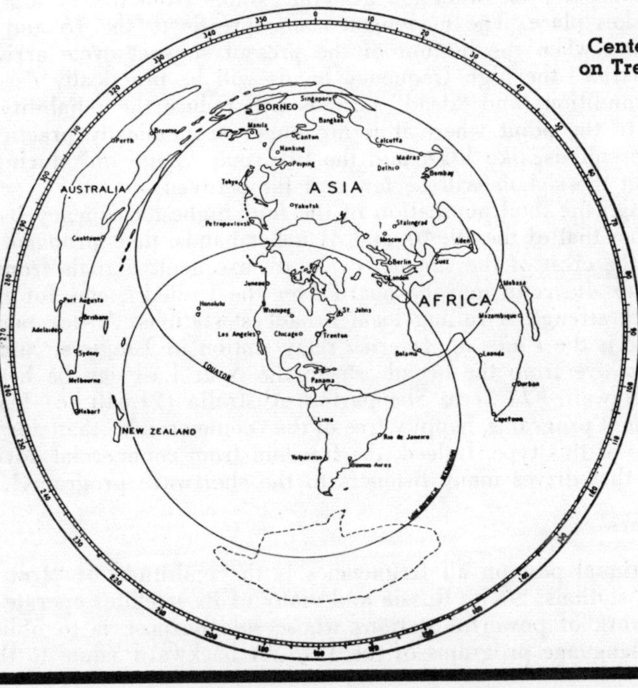

Centered on Trenton

will fade out and the listener will be treated to a few fascinating moments of entertainment from New Guinea, Cameroons, Cyprus, or some other far-off spot. *Radio Singapore* (7260 kc) and *The Voice of Righteousness,* Formosa (7300 kc) are both heard well in the Western part of the U.S., and the former is often heard in the Eastern part of the country.

Reception during the evening hours in the 41 meter band is chaotic, since it overlaps the 40 meter amateur phone band. The loud, local amateur signals obliterate all but the most powerful broadcast stations.

Popular inhabitants of the 31 meter band are the "down under" stations of Australia and New Zealand. Listen for the powerful signal of *Radio Australia* and the laugh of the Kookaburra bird, native to Australia. The song "Waltzing Matilda" signals the popular transmissions of this station. *Radio New Zealand* of Wellington features the call of the Bell Bird, while *Radio Sarawak* strums an identifying guitar.

The 25, 19, 16, 13, and 11 Meter Bands

These four bands, high in frequency, are considered seasonal bands. During periods of great sunspot activity, the ionosphere will support propagation at these frequencies. A steady migration of stations to the higher frequencies takes place over a period of years. As we are now just past the peak of the present cycle, these bands will support reliable communication for several more years. During the summer months the 11, 13, and 16 meter bands will be erratic, exhibiting their best characteristics during the fall and spring season. The 19 and 25 meter bands are all-year performers. As the sunspot cycle wanes, a gradual exodus from the 11 and 13 meter bands takes place. The movement soon spreads to the 16 and 19 meter bands, and when the bottom of the present sunspot cycle arrives in the early "sixties" the high frequency bands will be practically deserted. The erratic conditions and "dead" periods will reduce the reliability of transmission to the point where it is no longer economically practical. A few hardy perennials, like VOA and the BBC may "hang on" during the lean years, but the signals will be few and far between.

Although the total population of the four highest frequency bands never approaches that of the 49, 40, and 31 meter bands, their propagation conditions at the crest of the sunspot cycle are excellent. Signals from the BBC and *Radio Australia* may be heard over the United States for hours at a time, with strength rivalling local broadcast stations. A star performer at 21,455 kc is the *Voice of America* relay station in Tangiers, North Africa. The backwave from the broadcasts to the Near East can be heard as far west as Hawaii. *VLC21* in Shepparton, Australia (21,540 kc) features musical request programs, happily free of the "commercials" that clog American programs of this type. Indeed, the freedom from commercial advertising is a factor that drives many listeners to the shortwave programs!

"Jammers"

A continual pest on all frequencies is the multitude of "Iron Curtain" jamming stations. Soviet Russia and many of its satellites operate an extensive network of powerful stations whose sole purpose is to obliterate the Russian language programs of the VOA. A backward salute to the success

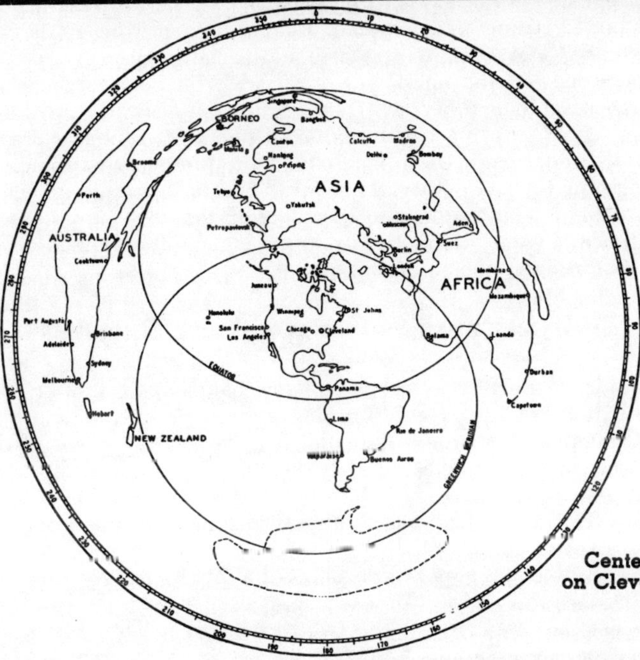

Centered on Cleveland

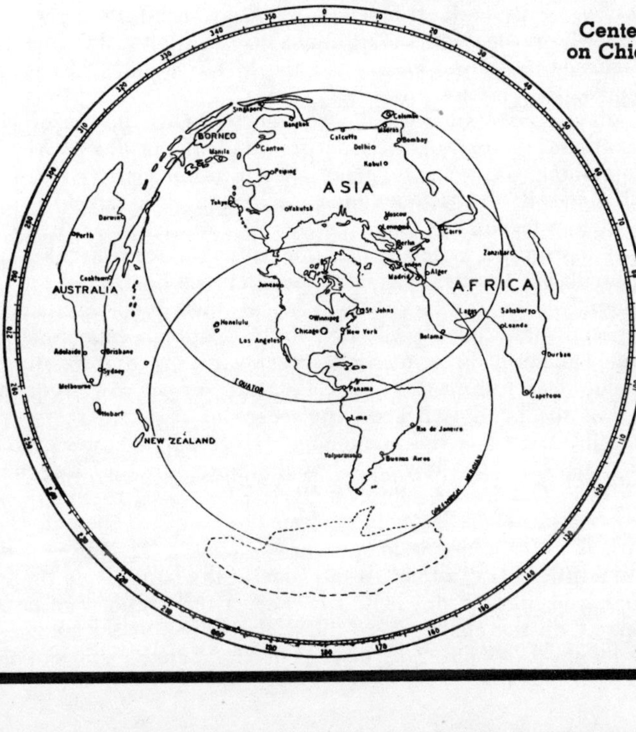

Centered on Chicago

of these programs is the swarm of jammers that lie athwart the broadcast bands, trying to strangle free speech. At one moment the air is clear, with no interference. The VOA will start a native language broadcast to some Soviet Republic, and the band will erupt with buzz-saw signals. As many as 25 jammers at one time have been counted in the 21 mc band. Each jammer has an identifying coded signal of one or two characters. With an atlas and a little imagination you can soon pin-point a number of the furry things on the map. Pity the poor SWL or ham living near one of these creatures! It is interesting to speculate on the effectiveness of this kind of interference. Usually, it is merely annoying, as the trained listener can often "hear through" the jamming signal.

A Bandspread Scale for Your Receiver

Many modern shortwave receivers have bandspread tuning dials. The bandspread dial provides vernier tuning action, permitting more accurate tuning and logging of stations. This dial takes a small segment of the shortwave spectrum and "magnifies" it over the entire rotation of the bandspread scale. Most dials of this sort are calibrated "0-100", and a few are calibrated in kilocycles for the amateur bands. Practically none of them are calibrated for the shortwave broadcast bands.

With a sheet of graph paper, a pencil, and a few hours' listening time, you can make calibration curves for each shortwave broadcast band, as interpreted against the markings of the bandspread dial. This curve will assist you in hunting down stations known to be active in the region covered by the calibration. With care, frequency readings accurate to two or three kilocycles may be made. You should make curves for the 11, 13, 16, 19 and 25 meter bands. Calibration curves for the low frequency bands often will come in handy, too. Here's how:

First of all you need an accurate frequency list of the major shortwave broadcast stations. An excellent list is provided in the *World Radio Handbook*. Let's assume you wish to make a calibration chart covering the 13 meter band. Accurate frequencies must be logged in kilocycles rather than meters, so we'll refer to this band as the 21,450-21,750 kc band, and we will calibrate our graph in terms of kilocycles. Make up a chart as shown in Figure 5, with the frequency range placed on the horizontal axis, and the bandspread dial range marks on the vertical axis. We'll also assume that your bandspread dial reads "100" with the bandspread capacitor open, and "0" with the capacitor fully meshed. Start with the dial reading of "0', and adjust the main tuning dial until a station on the low frequency edge (21,450 kc) of the band is heard. The *Voice of America* relay station in Tangier, North Africa is a fine band-edge marker. It operates on 21,455 kc (21.455 mc), right on the "low edge" of the band. Another fine marker for you is *WLWO* in Cincinnati, Ohio, at 21,485 kc. Now, readjust your dials so that the *VOA* station falls about "5" on the bandspread dial. If you are working with *WLWO,* place it about "12½" on the bandspread dial. This will give you a little extra "room" at the edge of the band. Note the setting of the main dial, and mark it down in a corner of the graph. The next step is to mark a point on the chart where the dial reading of the marker station and the frequency of the station meet. For the Tangier station, the coor-

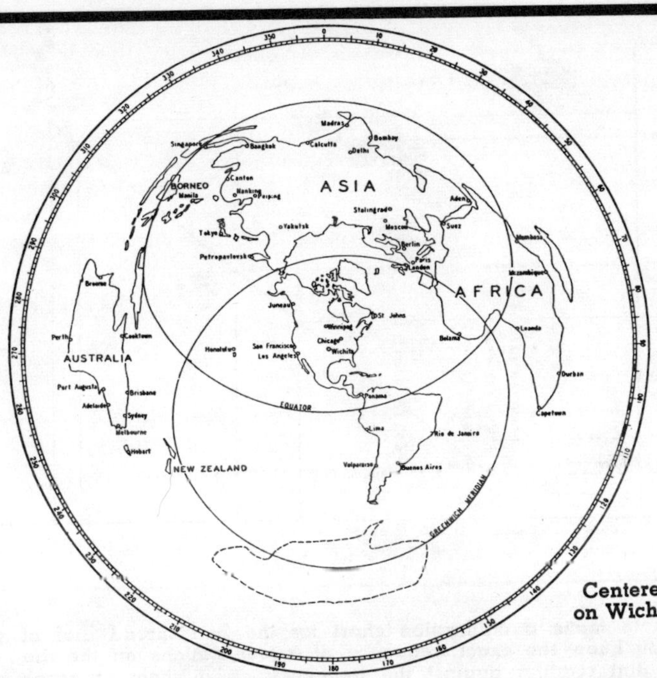

Centered on Wichita

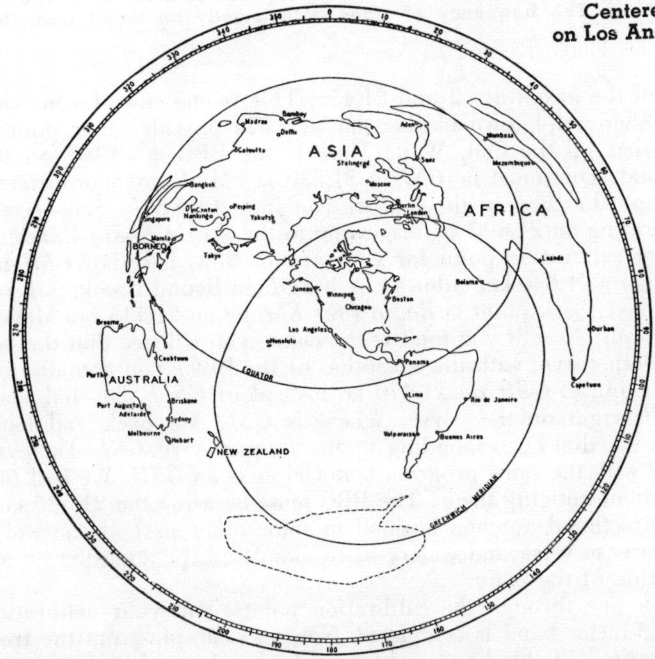

Centered on Los Angeles

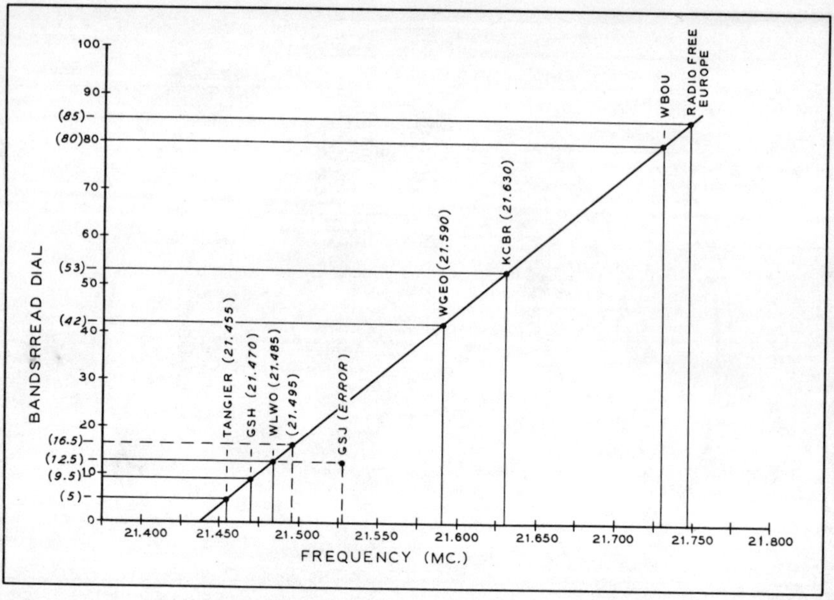

It's easy to make a calibration chart for the bandspread dial of your set! When you know the exact frequency of a few stations on the dial, you can plot the dial reading against the frequency on a sheet of graph paper as shown above. The frequency of other stations may be found from the graph.

dinates of the graph are 2 and 21,455. This is one end of your calibration curve. Other well-known marker stations will provide other points. Let us look further up the dial. Wait! Here is the BBC at "9½" on the bandspread dial. This must be *GSJ* at 21,530 kc. Mark an appropriate spot on your graph. Farther up the dial we run into *WGEO* in Schenectady, N.Y. It is operating on 21,590 kc, according to the *World Radio Handbook*. This is another calibration point for your graph. Now, here is *KCBR* in Dixon, California on 21,630 kc, fallowed by *WBOU* in Bound Brook, N.J. on 21,730 kc. The next check point is *Radio Free Europe* on 21,745 kc. Mark it down on the graph! Now if you look at the chart, you will see that the points fall on a smooth curve, with the exception of the BBC. You actually must have been listening to *GSH* on 21,470 kc instead of *GSJ!* The dial reading for *GSH* falls right on the curve. Where is *GSJ?* Go back and look at the point on the dial corresponding to the frequency of *GSJ*. There they are, about S2 with the same program material as is on *GSH*. We had tuned past them without noticing them! The BBC must be using the 21,530 kc channel with a directional antenna beamed to some other part of the world, since the signal is so weak and watery as to almost escape attention on a cursory examination of the band.

Draw a line through the calibration points, and your calibration curve for the 13 meter band is completed. Now you can pin-point the frequencies of any stations in this band right on your bandspread dial. If you wish to

CALL	LOCATION	FREQ. (MC.)	TIME TICK	MODULATION ∿	POWER (KW)
LOL	BUENOS AIRES, ARGEN.	2.5, 5, 10, 15, 20, 25	YES	440 1000	2.0
ZUO	JOHANNESBURG, UNION OF SO. AFRICA	5.0	YES	—	0.1
—	MOSCOW, USSR	10, 15	YES	—	—
MSF	RUGBY, ENGLAND	2.5, 5.0, 10	YES	1000	0.5
JJY	TOKYO, JAPAN	2.5, 4, 5, 8, 10, 15	YES	1000	1.0
IBF	TORINO, ITALY	5.0	YES	440 1000	0.3
—	UCCLE, BELGIUM	2.5	—	—	0.02
CHU	OTTAWA, CANADA	3.330, 7.335, 14.76	—	1000	3.0
WWV	BELTSVILLE, MD.	2.5, 5, 10, 15, 20, 25	YES	440 1000	1-9
WWVH	PUUNEME, T.H.	5, 10, 15			

WWV–WWVH PROPAGATION FORECASTS

LETTER = *QUALITY OF PROPAGATION AT TIME OF BROADCAST.*

W – DISTURBED
U – UNSETTLED
N – NORMAL

DIGIT = *QUALITY OF FORECAST FOR NORTH ATLANTIC (WWV) OR NORTH PACIFIC (WWVH) PATH DURING NEXT 6 HOURS.*

1 – USELESS
2 – VERY POOR
3 – POOR
4 – POOR TO FAIR
5 – FAIR
6 – FAIR TO GOOD
7 – GOOD
8 – VERY GOOD
9 – EXCELLENT

Fig. 4 Standard frequency signals are broadcast from many points on the globe. WWV and WWVH blanket the Northern Hemisphere, and the other signals may be heard in the U.S. Look for JJY on the "top edge" of the 80-meter amateur phone band during the winter mornings. CHU is easy to find, also.

find *Radio Nacional* in Portugal on 21,495 kc, you will tune to "16½", and that's where they will be, *provided* you always re-set the main tuning dial so that the known "marker" stations fall at the proper bandspread dial calibration points.

This system is very handy, but it is not absolutely foolproof, as some of the s-w broadcast stations have a disconcerting habit of hopping and skipping about on the dial as they try to find an interference-free channel. Some of the smaller stations, too, are not always on frequency, drifting about slowly during their programs. However, once you set your dials correctly and know the approximate frequency of your quarry, you can hunt it down like a bloodhound! Make up a set of charts, one for each shortwave broadcast band, and you will find that they will help your DX-efforts a thousand percent!

You can make up a set of bandspread calibration charts for each amateur band, if you wish. It is a little harder to obtain calibration points in the ham bands, as the amateurs do not usually announce their operating frequency, and they hop about the band with more gusto than do the s-w broadcasting stations. Even so, the high frequency and low frequency edges of the phone band may be approximated at the point where the din drops off, and the Canadian and foreign amateurs appear. The same approximation my be mde for the edges of the c-w bands where an abrupt cessation of c-w activity is noted. After several listening periods you will be able to accurately mark the band edges on your chart.

INDEX